# The
# Global Positioning System
# and GIS

# The

# Global Positioning System

# and GIS

**Michael Kennedy**
University of Kentucky

**Ann Arbor Press, Inc.**
Chelsea, Michigan

**Library of Congress Cataloging-in-Publication Data**
Catalog record is available from the Library of Congress

ISBN 1-57504-017-4

ANN ARBOR PRESS, INC.
121 South Main Street, Chelsea, Michigan 48118

PRINTED IN THE UNITED STATES OF AMERICA
1 2 3 4 5 6 7 8 9 0

For

Alexander Kennedy

# About the Author

Michael Kennedy's involvement with Geographic Information Systems began in the early 1970s with his participation on a task force formed by the Department of the Interior to provide technical recommendations for pending federal land use legislation.

In the mid-1970s he and co-authors wrote two short books on GIS, both published by the Urban Studies Center at the University of Louisville where he was enjoying sabbatical leave from the University of Kentucky. *Spatial Information Systems: An Introduction* with Charles R. Meyers is a description of the components of a GIS and was a guide to building one at the time when there was no off-the-shelf software. *Avoiding System Failure: Approaches to Integrity and Utility* with Charles Guinn, described potential pitfalls in the development of a GIS. With Mr. Meyers and R. Neil Sampson, he also wrote the chapter "Information Systems for Land Use Planning" for *Planning the Uses and Management of Land*, a monograph published in 1979 by the American Society of Agronomy.

Professor Kennedy is also a computer textbook author, having co-written, with Martin B. Solomon, *Ten Statement Fortran Plus Fortran IV*, *Structured PL/ZERO Plus PL/ONE*, and *Program Development with TIPS and Standard Pascal*, all published by Prentice-Hall,

Over the years Professor Kennedy has had a wide range of experiences relating computers and environmental matters. Primarily to be able to talk to planners about the newly emerging field of GIS he became certified as a planner by the American Institute of Certified Planners (AICP). He was Director of the Computer-Aided Design Laboratory at the University of Kentucky for several years. He has been invited to teach GIS and/or programming at Simon Fraser University and several state or provincial universities: North Carolina, Florida, and British Columbia.

Outside of his interest in the Global Positioning System (GPS) the author's primary concern is in the development of computer data structures for the storage of geographic information. In work sponsored by the Ohio State University Center for Mapping and the Environmental Systems Research Institute, he is currently developing what he calls the dot-probability paradigm for the storage of spatial data. Fundamentally, the author is a programmer who has sought out the application of computers to environmental issues. He is currently an Associate Professor in the Geography Department at the University of Kentucky, where he teaches GIS.

# Contents

## Part 3 — Examining GPS Data

## Part 4 — Differential Correction

## Part 5 — Integrating GPS Data with ARC/INFO

## Part 6 — ArcView, ArcData, and GPS

## Part 7 — The Present and the Future

# Foreword

Michael Kennedy's latest book brings together Geographic Information System (GIS) technology and Global Positioning System (GPS) technology with the aim of teaching how to couple them to effectively capture GPS data in the field and channel it to a GIS.

We at ESRI were especially pleased that he chose to use ESRI's GIS software (ARC/INFO and ArcView) in writing his book, and we were happy to be able to provide him with some support in his efforts.

After fifteen or twenty years in which very few textbooks were written about GIS and related technologies there is now a veritable flood of new GIS books coming into print. Why is this one especially valuable?

First, because it couples GPS/GIS in an especially intimate way. Michael's intention in writing was to make it possible for readers working alone or for students in a formal course to learn how to use GPS/GIS "hands on"; to walk away from this textbook ready to go into the field and start using Trimble Navigation's GeoExplorer and ESRI's ARC/INFO and ArcView software to collect GPS field data and enter it into a GIS for immediate use.

Besides providing step-by-step instructions on how to do this, he provides appropriate background information in the form of theoretical discussions of the two technologies and examples of their use. He writes in an easy style, explaining the needed technical and scientific principles as he goes, and assuming little in the way of necessary prior instruction.

Instructors will find the text especially useful because Michael provides the kinds of detailed procedures and hints that help to make lab work "bullet proof" even for the inexperienced student. An accompanying CD-ROM has data which are likely to be useful in various ways to teacher and student alike.

Secondly, this book is important because GPS/GIS is such an extremely important technology! It is no exaggeration to say that GPS/GIS is *revolutionizing* aspects of many fields, including surveying (slashing the costs of many kinds of survey efforts and bringing surveying to parts of the world where surveys are non-existent, highly inaccurate, or long since outdated), the natural resource fields (providing rapid and far more accurate collection of field natural resource data of many kinds), and municipal planning (providing for the updating of all kinds of records based on accurate field checking), to name only a few. GPS is making practical the kinds of data collection which were simply out of the question only a few years ago because the necessary skilled teams of field personnel were unavailable and the costs of accurate field data collection were beyond the means of virtually all organizations which needed these kinds of data. GPS/GIS is changing all that. Use of the kinds of methods taught in *The Global Positioning System and GIS* is spreading very rapidly; GPS/GIS use will become commonplace throughout dozens of fields in just the next few years as costs of hardware and software continue to fall and books like this one increase the number of persons familiar with these two coupled technologies.

The impact of this revolution in data gathering will, I believe, have profound effects on the way in which we view the earth, on ways in which we exercise our stewardship of its resources for those who come after us, and on the professional practice of an extraordinary range of disciplines (engineering, oceanography, geology, urban planning, archaeology, agriculture, range management, environmental protection, and many, many others). I look forward to the time when *tens of millions of people* will make use of GPS/GIS technology every day, for thousands of purposes.

ESRI's aim as a company has always been to provide reliable and powerful GIS and related technologies to our clients and users and to help them use these technologies to make their work more effective and successful. By doing so, we hope to help make a difference in the world.

The ESRI authors program was created to further those same goals. Michael Kennedy's *The Global Positioning System and GIS* is an extremely important part of that program and will, I believe, assist many persons to acquire and effectively use GPS/GIS in their work. In writing this book he has performed an important service, not just to his readers and the users of this textbook, but to those whose lives will be improved because of the use of GPS/GIS technologies in the years ahead.

Jack Dangermond
President, ESRI, Inc.

# Preface for the Instructor[1]

## Purpose and Audience

The purpose of this textbook/workbook, and the accompanying CD-ROM, is to provide a short, intermediate, or full term course in using the Global Positioning System (GPS) as a method of data input to a Geographic Information System (GIS). The short course may either "stand alone" or be a two to three week segment in a general course in GIS. The text may either be used in a formal course with several students or as an individual self-teaching guide. There is the assumption that the students have at least a passing familiarity with GIS and with the most basic geographical concepts, such as latitude and longitude. But the book can serve as a gentle introduction to GPS even for those who do not intend to use the data in a GIS.

## A Look at the Contents

The text is divided into seven Parts:

Part 1 – Basic Concepts – is an introduction to GPS as a system. The field/lab work involves data collection with pencil and paper.

Part 2 – Automated Data Collection – looks further into how GPS works. Project work includes taking GPS data and storing it in files, first in a receiver, then in a PC.

Part 3 – Examining GPS Data – first answers some questions to complete the discussion of the theory of GPS, and then gives the student, or other reader, experience with software for processing and displaying collected GPS data.

---

[1] Material for the student begins with the Introduction on page 1.

Part 4 – Differential Correction – discusses the issue of accuracy and techniques for obtaining it. Practice first takes place with "canned" data – then with the user's own.

Part 5 – Integrating GPS and GIS Data – has the reader putting GPS data into the ARC/INFO GIS. The issues of positional correctness are again addressed.

Part 6 – ArcView, ArcData, and GPS – is a discussion of ESRI's ArcView product and an extensive exercise that is based on data from both GPS and the ArcUSA database.

Part 7 – The Present and the Future – discusses the state of practice regarding several matters: collecting feature attribute data, navigating with GPS, real-time differential GPS, and mission planning, It provides for hands-on exercises relating to these topics. Here also are some guesses about what is to come.


**Approach: Both Theoretical and Hands-on**

The approach I use is to divide each Part into two modules:

- OVERVIEW
- STEP-BY-STEP

This division comes from my belief that learning a technical subject such as GPS involves two functions: education and training. The student must understand some of the theory and the language of the subject he or she is undertaking; but also, the ability to perform tasks using the technology is likewise important, and such hands-on experience provides new insights into the subject. Many texts attempt to perform both of these functions, but mix the relevant text together. It is my view — particularly in the case of GPS which involves several complex systems, field work, and learning about hardware and software — that the two functions serve the learner best if they are distinctly separated. The theory is presented in the "Overview" module. It is laid out in a hierarchical, simple-to-complex, way (from the top, down). The training portion takes place in a linear (step-by-step) form: "do this, now do this".

There is heavy emphasis on getting the vital parameters, such as datum and coordinate system, correct. No conscientious student should be able to leave a course based on this text and commit the all-too-easy sin of generating incorrect GIS information based on incorrectly converted GPS data.

## Hardware and Software

Teaching the U.S. Global Positioning System (NAVSTAR) using GIS with a "hands-on" approach involves using particular hardware and software. I decided it was not possible to write a satisfactory text which was general with respect to the wide variety of products on the market. I therefore selected the two of the most popular and capable systems available: the Trimble Navigation GeoExplorer GPS receiver,[2] and ARC/INFO and Arc/View GIS software systems from Environmental Systems Research Institute (ESRI). If you choose to use other products — and there are several fine ones in both GPS and GIS — you may still use the text, but you will have to be somewhat creative in applying the STEP-BY-STEP sections.

## An Instructor's Guide: On the CD ROM

An instructor who undertakes to teach a GPS/GIS course for the first time may face a daunting task, as I know only too well. In addition to many of the problems that face those teaching combination-lecture-and-lab courses, there are logistical problems created by the (assumed) scarcity of equipment, the need to electrically charge the receivers, the outdoors nature of some of the lab work, and other factors. I have made an attempt keep other

---

[2] Included on the CD-ROM are sections that will allow you to use this text with Trimble Navigation's Pathfinder Basic receivers.

teachers from encountering the difficulties I found, and to anticipate some others. For example, paper forms are provided (both in the text and on the CD-ROM) to provide for equipment checkout and setup, and to provide base station information.

On the CD-ROM you will find outlines for courses of varying lengths and hands-on involvement levels.

## Demonstration Data

The text is quite flexible with respect to use of data supplied on the CD-ROM in conjunction with the data the students, usually working in pairs, collect on their own. Most assignments begin with manipulating the canned data, to prevent surprises, followed by students using home-grown data, if time and interest permit.

The step-by-step procedures have been tested, and tested again, so the projects and exercises should work as indicated — subject, of course, to new releases of software, firmware, and hardware

# Acknowledgements

The project that led to this text began in a conversation with Jack Dangermond in May of 1993. I proposed to quickly put together a short introduction to GPS for GIS users. Since that time I have learned a great deal about both GPS and "quickly putting together a short introduction." Several software releases later, several introductions of GPS hardware later, and many occasions on which I learned that there was a lot more to this subject than I imagined, it is finished. I could not have done it without:

- My daughter Heather Kennedy for considerable help and support along the way and, particularly, for the painstaking and intense work of final proofing and testing.
- My son Evan Kennedy, who added to the collection of GPS tracks presented in the text, by taking a GPS receiver across the U.S. by automobile, and whose enthusiasm for GPS encouraged me to complete the book.
- Allan Hetzel, who took on the job of fixing and printing the camera-ready copy. He coped with corrections from several reviewers and coordinated the final marathon production session.
- Dick Gilbreath, who, with the help and forbearance of Donna Gilbreath, spent many hours at inconvenient times producing most of the figures, and who insisted on getting the smallest details right.
- Yu Luo and Pricilla Gotsick of Morehead State University, who "burned" the CD-ROM used for the data.
- People at ESRI: Jack Dangermond who supported the idea of this text. Bill Miller, Earl Nordstrand, and Michael Phoenix who provided encouragement and advice.
- People at Trimble Navigation: Art Lange, my GPS guru, who provided considerable technical help and kept me from making several real blunders. Chuck Gilbert, Chris Ralston, and Dana Woodward, who helped by providing advice and equipment. Michele Vasquez, who provided photos for the text.

- Carla Koford and Ethan Bond, of the GIS lab at the University of Kentucky (UK) who tested and corrected procedures and text, and Jena King, who read and improved parts of the text.
- Justin Stodola, who wrote the "C" program and the procedures for digitizing coverages directly into unique UTM tiles.
- People who collected GPS data in faraway places: Will Holmes for the Mexican data and Chad Staddon, who took a GPS receiver to Bulgaria.
- Bob Crovo, of the UK computing center, who was always cheerful about answering dumb questions.
- Ron Householder of MapSync, and Timouthy Poindexter of CDP Engineers, who use GPS as professionals and know a lot that isn't in the manuals.
- David Lucas, GIS coordinator for Lexington–Fayette County, who guided me through some sticky problems with UNIX and license managers and provided data on Lexington Roads.
- Ken Bates — Mr. GIS for the state government of Kentucky, and Kent Annis of the Bluegrass Area Development District, for their help and insights.
- The students in several classes of GEO 409, 506, and 509 — GIS and computer-assisted cartography courses at UK — who read the book and tested the exercises.
- Ruth Rowles, who used an early version of the text in her GIS class at UK.
- Calvin Liu, who operates the GPS community base station at UK and provided many of the base station files used herein. And the folks at the base station in Whiterock, B.C. for helping a stranger with an urgent request.
- Scott Samson, who provided good advice at important times.
- Jon Goss and Matt McGranaghan facilitated my stay and lecture at the University of Hawaii — one of the nicer places to collect GPS data, or data of any sort for that matter.
- Tom Poiker, of Simon Fraser University, for his help and counsel on various GIS topics, and to him and Jutta for

their hospitality in their home in British Columbia — also a great place for data collection.

- Max Huff, of OMNISTAR, Inc. who demystified the complex "differential corrections anywhere" system.
- Several companies for hardware, software, and support: ESRI for sponsoring the project and providing software; Trimble Navigation, for providing the GPS hardware; AccuPoint and OMNISTAR for access to their differential correction services.
- John Bossler, of Ohio State University, who does really hi-tech GPS, and who was patient in letting me finish this text, delaying a project I am doing for him.
- To the folks at Ann Arbor Press, who did indeed "press" to get the book into final form: Skip DeWall and Sharon Ray. I should add that responsibility for the appearance of the text and any errors you may find rest entirely with the author. While I received considerable help in development of the material, I provided the camera-ready text and was responsible for the correctness and format of the final work.
- To my colleagues in the Department of Geography at the UK — and particularly to Richard Ulack, its chair — who indulged my absences and absent-mindedness during the last weeks of this project.
- And finally to my dear friend Barb Emler, who repairs children's teeth from nine to five every day and justly believes that people ought to enjoy life without work outside of those times. She's right. And I will. For a while, anyway.

# Introduction

Two of the most exciting and effective technical developments to emerge in the last decade are:

- the introduction of Global Positioning Systems (the GPS of the United States is called NAVSTAR), and
- the phenomenon of the Geographic Information System (GIS).

GIS is an extremely broad and complex field, concerned with the use of computers to input, store, retrieve, analyze, and display geographic information. Basically GIS programs make a computer think it's a map — a map with wonderful powers to process spatial information, and to tell its users about any part of the world, at almost any level of detail.

While GPS is also an extremely complex system, using it is simple by comparison. It allows you to know where you are by consulting a radio receiver. The accuracies range from as good as a few millimeters to somewhere around 100 meters, depending on equipment and procedures applied to the process of data collection.

More advanced GPS receivers can also record location data for transfer to computer memory, so GPS can not only tell you where you are — but also tell you where you were. Thus, GPS can serve as means of data input for GISs. Traditionally (if one can use that word for such a new and fast moving technology), GISs got their data from maps and aerial photos. These were either scanned by some automated means or, more usually, digitized manually using a hand-held "puck" to trace map features — the map being placed on an electronic drafting board called a "digitizer." With GPS, the earth's surface becomes the digitizer board; the GPS receiver becomes the puck. This approach inverts the entire traditional process of GIS data collection: spatial data come directly from the environment and the map becomes a document of output rather than input.

The aim of this text is to teach you to use GPS as a source of input to GIS.

# Part 1

# Basic Concepts

IN WHICH *you are introduced to facts and concepts relating to the NAVSTAR Global Positioning System and have your first experience with a GPS receiver.*

# Basic Concepts

## OVERVIEW

*A sports club in Seattle decided to mount a hunting expedition. They employed a guide who came well recommended, and whose own views of his abilities were greater still. Unfortunately, after two days, the group was completely, totally lost. "You told me you were the best guide in the State of Washington," fumed the person responsible for hiring the guide. "I am, I am," claimed the man defensively. "But at the moment I think we're in Canada."*

Stories like the one above should be told now (if at all), before they cease to be plausible. Actually, even at present, given the right equipment and a map of the general area, you could be led blindfolded to any spot in the great out-of-doors and determine exactly where you were. This happy capability is due to some ingenious electronics and a dozen billion dollars spent by the U.S. government. I refer to NAVSTAR (NAVigation System with Time And Ranging) — a constellation of 24 satellites orbiting the

Earth, broadcasting data that allows users on or near the Earth to determine their spatial positions. The general term for such an entity is "Global Positioning System" or "GPS." The Russians have a GPS as well, which they call GLONASS. (One might reflect that, for some purposes, the cold war lasted just long enough.) In the western world, GPS usually implies NAVSTAR, so I will use the two designations interchangeably in this text.

## Where Are You?

Geography, and GIS particularly, depend on the concept of location. Working with "location" seems to imply that we must organize and index space. How do we do that?

Formally, we usually delineate geographical space in two dimensions on the Earth's surface with the latitude-longitude graticule, or with some other system based on that graticule.

But informally, and in the vast majority of instances, we organize space in terms of the features in that space. We find a given feature or area based on our knowledge of other features — whether we are driving to Vancouver or walking to the refrigerator. Even planes and ships using radio navigational devices determine their positions relative to the locations of fixed antennae (though some of the radio signals may be converted to graticule coordinates).

Unlike keeping track of *time*, which was initially computed relative to a single, space-based object (the sun), humans kept track of *space* — found their way on the ground — by observing what was around them.

Another, somewhat parallel way of looking at this issue is in terms of absolute versus relative coordinates. If I tell you that Lexington, Kentucky is at 38 degrees (38°) north latitude, 84.5° west longitude, I am providing you with absolute coordinates. If I say, rather, that Lexington is 75 miles south of Cincinnati, Ohio and 70 miles east of Louisville, Kentucky, I have given you relative coordinates.

Although relative coordinates can be quite precise, they usually appeal more to our intuitive comprehension of "location" than do absolute coordinates.

To pass spatial information around, humans developed maps to depict mountains and roads, cities and plains, radio stations and sink holes. Maps aid both the formal and informal approaches that humans use to find objects and paths. Some maps have formal coordinates, but maps without graticule markings are common. All maps appeal to our intuitive sense of spatial relationships. The cartographer usually relies on our ability to use the intuitive coordinates in our memory, and our abilities to analyze, to extrapolate, and to "pattern match" the features on the map. It is good that this method works, since, unlike some amazing bird and butterfly species, humans have no demonstrated sense of an absolute coordinate system. But with maps, and another technological innovation, the magnetic compass, we have made considerable progress in locating ourselves.

I do not want to imply that absolute coordinates have not played a significant part in our position-finding activities. They have, particularly in navigation. At sea, or flying over unlit bodies of land at night, captains and pilots used methods that provided absolute coordinates. One's position, within a few miles, can be found by "shooting the stars" for a short time with devices such as sextants or octants. So the GPS concept — finding an earthly position from bodies in space — is not an entirely new idea. But the ability to do so during the day, almost regardless of weather, with high accuracy and almost instantaneously, makes a major qualitative difference. As a parallel, consider that a human can move by foot or by jet plane. They are both methods of locomotion, but there the similarity ends.

GPS, then, gives people an easy method for both assigning and using absolute coordinates. Now, humans can know their **positions** (i.e., the coordinates that specify where they are); combined with map and/or GIS data they can know their **locations** (i.e., where they are with respect to objects around them). I hope that, by the time you've completed this text and experimented with a GPS receiver, you will agree that NAVSTAR constitutes an astounding leap forward.

## GPS and GIS

The subject of this book is the use of GPS as a method of collecting locational data for Geographic Information Systems (GIS). The appropriateness of this seems obvious, but let's explore some of the main reasons for making GPS a primary source of data for GIS:

- Availability: In 1995, the U.S. Department of Defense declared NAVSTAR to have "final operational capability." Deciphered, this means that the U.S. DoD has committed itself to maintaining NAVSTAR's capability for civilians at a level specified by law, for the foreseeable future, at least in times of peace. Therefore, those with GPS receivers may locate their positions anywhere on the Earth.
- Accuracy: GPS allows the user to know position information with remarkable accuracy. At least two factors promote such accuracy:

  First, with GPS, we work with primary data sources. Consider one alternative to using GPS to generate spatial data: the digitizer. A digitizer is essentially an electronic drawing table, wherein an operator traces lines or enters points by "pointing" — with "cross-hairs" embedded in a clear plastic "puck" — at features on a map.

  One could consider that the ground-based portion of a GPS system and a digitizer are analogous: the Earth's surface is the digitizer tablet, and the GPS receiver antenna plays the part of the cross-hairs, tracing along, for example, a road. But data generation with GPS takes place by recording the position on the most fundamental entity available: the Earth itself, rather than a map or photograph of a part of the Earth that was derived through a process involving perhaps several transformations.

  Second, GPS itself has high inherent accuracy. The precision of a digitizer may be 0.1 millimeters (mm). On a map of scale 1:24,000, this translates into 2.4

meters (m). A distance of 2.4 m is comparable to the accuracy one might expect of the properly corrected data from a medium-quality GPS receiver. It would be hard to get this out of the digitizing process. A secondary road on our map might be represented by a line five times as wide as the precision of the digitizer (0.5 mm wide), giving a distance on the ground of 12 m, or about 40 feet.

On larger-scale maps, of course, the digitizer precision can exceed that obtained from the sort of GPS receiver commonly used to put data into a GIS. On a "200 scale map" (where one inch is equivalent to 200 feet on the ground) 0.1 mm would imply a distance of approximately a quarter of a meter, or less than a foot. While this distance is well within the range of GPS capability, the equipment to obtain such accuracy is expensive and is usually used for surveying, rather than for general GIS activities. Finally, at the extremes of accuracy, GPS wins over all other methods. GPS can provide horizontal accuracies of less than one centimeter.

- Ease of use: Anyone who can read coordinates and find the corresponding position on a map can use a GPS receiver. A single position so derived is accurate to within a hundred meters or so. Those who want to collect data accurate enough for a GIS must involve themselves in more complex procedures, but the task is no more difficult than many GIS operations.

- GPS data are inherently three-dimensional: In addition to providing latitude-longitude (or other "horizontal" information), a GPS receiver may also provide altitude information. In fact, unless it does provide altitude information itself, it must be told its altitude in order to know where it is in a horizontal plane. The accuracy of the third dimension of GPS data is not as great, usually, as the horizontal accuracies. As a rule of thumb, variances in the horizontal accuracy should be multiplied by two (and perhaps as much as five) to get an estimate of the vertical accuracy.

## Anatomy of the Term: "Global Positioning System"

**Global:** Anywhere on Earth. Well, almost anywhere, but not inside buildings, underground, in severe precipitation, under heavy tree canopy, or anywhere else not having a direct view of a substantial portion of the sky. The radio waves that GPS satellites transmit have very short lengths — about 20 cm. A wave of this length is good for measuring because it follows a very straight path, unlike its longer cousins, such as AM and FM band radio waves, that may bend considerably. Unfortunately, short waves also do not penetrate matter very well, so the transmitter and the receiver must not have much solid matter between them, or the waves are blocked, as light waves are easily blocked.

**Positioning:** Answering brand new and age-old human questions. Where are you? How fast are you moving and in what direction? What direction should you go to get to some other specific location, and how long would it take at your speed to get there? *And*, most importantly for GIS, where have you been?

**System:** A collection of components with connections (links) among them. Components and links have characteristics. GPS might be divided up in the following way:

The Earth

The first major component of GPS is Earth itself: its mass and its surface, and the space immediately above. The mass of the Earth holds the satellites in orbit. From the point of view of physics, each satellite is trying to fly by the Earth at four kilometers per second. The Earth's gravity pulls on the satellite vertically so it falls. The trajectory of its fall is a track that is parallel to the curve of the Earth's surface.

The surface of the Earth is studded with little "**monuments**" whose positions are known quite accurately. These lie in the "numerical **graticule**" which we all agree forms the basis for geographic position. Measurements in the units of the graticule, and based on the positions of the monuments, allow us to determine the position of any object we choose on the surface of the Earth.

## Earth-circling satellites

The U.S. GPS design calls for a total of 24 solar-powered radio transmitters, forming a constellation such that several are "visible" from any point on Earth at any given time. The first one was launched on February 22, 1978. In mid-1994 all 24 were broadcasting. The standard "constellation" of 24 includes three "spares".

The satellites are at a "middle altitude" of 20,200 kilometers (km), roughly 12,600 statute miles or 10,900 nautical miles (nm), above the Earth's surface. This puts them above the standard orbital height of the space shuttle, most other satellites, and the enormous amount of space junk that has accumulated. They are also well above Earth's air, where they are safe from the effects of atmospheric drag. GPS satellites are below the geosynchronous satellites, usually used for communications and sending TV and other signals back to Earth-based fixed antennas. These satellites are 35,420 km (or 19,130 nm) above the Earth, where they hang over the equator relaying signals from and to ground-based stations.

The NAVSTAR satellites are neither polar nor equatorial, but slice the Earth's latitudes at about 55°, executing a single revolution every 12 hours. Further, although each satellite is in a 12 hour orbit, an observer on Earth will see it rise and set about 4 minutes earlier each day.[1] There are four satellites in each of six distinct orbital planes. The orbits are almost exactly circular. The combination of the Earth's rotational speed and the satellites' orbits produces a wide variety of tracks across the Earth's surface. Below is a view of the tracks which occurred during the first two hours after noon on St. Patrick's Day, 1996. You are looking down on the Earth, directly at the equator; the north-south meridian

---

[1] Why? Answers to several questions that may occur to you will be supplied in Part 2. We avoid the digression here.

passes through Lexington, Kentucky. As you can see, the tracks near the equator tend to be almost north-south.

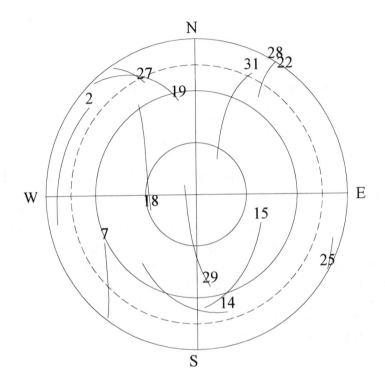

Fig. 1-1 — GPS Satellite tracks looking toward the equator

GPS satellites move at a speed of 3.87 km/sec (8,653 miles per hour). Each weighs about 860 kilograms (roughly a ton) and has a size of about 8.7 meters (about 17 feet) with the solar panels extended. Space buffs might want to know that they usually get into orbit on top of Delta II rockets fired from the Kennedy Spaceflight Center in Florida.

Fig. 1-2 — A NAVSTAR satellite

## Ground-based stations

While the GPS satellites are free from drag by the air, their tracks are influenced by the gravitational effects of the moon and sun, and by the solar wind. Further, they are crammed with electronics. Thus, both their tracks and their innards require monitoring. This is accomplished by four ground-based stations, located on Ascension Island, at Diego Garcia, in Hawaii, and Kwajalein. Each satellite passes over at least one monitoring station twice a day. Information developed by the monitoring station is transmitted back to the satellite, which in turn re-broadcasts it to GPS receivers. Subjects of a satellite's broadcast are the health of the satellite's electronics, how the track of the satellite varies from what is expected, the current **almanac**[2] for all the satellites, and other, more esoteric subjects which need not concern us at this point. Other ground-based stations exist, primarily for uploading information to the satellites; the master control station is in Colorado Springs, Colorado.

## Receivers

This is the part of the system with which you will become most familiar. In its most basic form, the satellite receiver consists of

- an antenna (whose position the receiver reports),
- electronics to receive the satellite signals,
- a microcomputer to process the data that determines the antenna position, and to record position values,
- controls to provide user input to the receiver, and
- a screen to display information.

More elaborate units have computer memory to store position data points and the velocity of the antenna. This information may be uploaded into a computer and then installed in a geographic information system. Another elaboration on the basic GPS unit is the ability to receive data from and transmit data to other GPS receivers — a technique called "real time differential GPS" that may be used to considerably increase the accuracy of position finding.

## Receiver manufacturers

In addition to being an engineering marvel and of great benefit to many concerned with spatial issues as complex as national defense or as mundane as re-finding a great fishing spot, GPS is also big business. Dozens of GPS receiver builders exist — from those

---

[2] An almanac is a description of the predicted positions of heavenly bodies.

who manufacture just the GPS "engine," to those who provide a complete unit for the end user.

## The United States Department of Defense

The U.S. DoD is charged by law with developing and maintaining NAVSTAR. It was, at first, secret. Five years elapsed from the first satellite launch in 1978 until news of GPS came out in 1983. In the decade since — despite the fact that parts of the system remain highly classified — mere citizens have been cashing in on what one manufacturer calls "The Next Utility." There is little question that the design of GPS would have been different had it been a civilian system "from the ground up." But then, GPS might not have been developed at all. Many issues must be resolved in the coming years. For example, the military deliberately corrupts the GPS signals so that a single GPS unit, operating by itself (i.e., **autonomously**), cannot assure accuracy of better than 100 meters. But the DoD is learning to play nicely with the civilian world. They and we all hope, of course, that the civil uses of GPS will vastly outpace the military need.

## Users

Finally, of course, the most important component of the system is you: the "youser," as an eight-year-old spelled it. A large and quickly growing population, users come with a wide variety of needs, applications, and ideas. From tracking ice floes near Alaska to digitizing highways in Ohio. From rescuing sailors to pinpointing toxic dump sites. From urban planning to forest management. From improving crop yields to laying pipelines. Welcome to the exciting world of GPS!

# STEP-BY-STEP

## PROJECT 1A

### Getting Acquainted with a GPS Receiver

You begin your first Global Positioning System project by becoming acquainted with a typical GPS receiver, while still inside a building. Your investigation begins with a Trimble GeoExplorer Receiver, and a notebook.

The GeoExplorer and the enhanced version, the GeoExplorer II, operate in much the same fashion, so I will simply use the name GeoExplorer to refer to both units. The primary external difference is that the "II" has a receptacle which allows use of an external antenna. The original receiver was built with only an internal antenna, making it less convenient in some situations — as inside an automobile, for instance.

The notebook — with paper and a sturdy writing surface — is there partly to get you used to the idea that a GPS receiver and a note taking ability must go hand-in-hand. After this first project, most of the data you take will be recorded in computer files, but some will not and must be written down. The notebook will also provide an index to the computer files you record.

A Trimble GeoExplorer receiver with built-in antenna is a hand-held device about 3.75" wide by 7" long and less than 2" thick, weighing less than a pound. It is usually powered in the field by a standard VideoCam battery which weighs about 1.5 pounds. Alternative power sources are "AA" batteries and the cigarette lighter receptacles in a car or boat.

While the unit is very complex internally, containing not only signal reception electronics, but a microcomputer with a quarter megabyte memory as well, its user controls are quite basic: eight buttons.

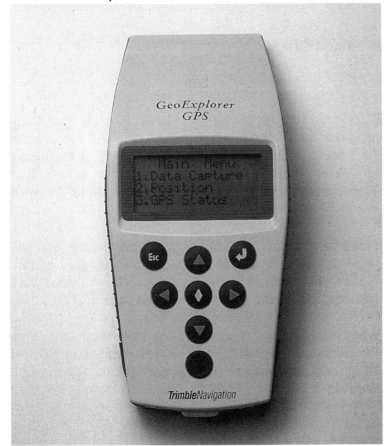

Fig. 1-3 — The GeoExplorer GPS receiver from Trimble Navigation

The button that is "furthest south" is the on-off control. The "up" and "down" buttons scroll through the lines of text and numbers of whatever screen is currently displayed. The "left" and "right" buttons move a cursor from character to character within a string of characters and perform other selection functions. The center button — the one with the diamond — is the "**Command**" key which sets options chosen by the others. We will designate this key as "**CMD**" from now on.

Visual output to the user is provided by a four-line, 16 character per line LCD display. In some instances the screen serves as a window to a list of more than four lines, but the user can see all of them by repeatedly pressing the "down" key (or the "up" key).

## Power On and Off

{__}[3]

{__} Find the cord with the small cylindrical jack on one end and the large cylinder on the other. Connect it to the receiver and the battery (or battery charger). A small green light at the base of the receiver should glow. The screen of the receiver may be blank, or may display characters. If you see characters, hold down the on-off button until the display becomes blank.

{__} Press the on-off button. Some preliminary screens will appear that describe the receiver type, firmware version, and other facts. Note these items in your notebook. (If you don't get all of them this time, check the next time you turn the unit on.)

{__} What you see now depends on the state the receiver was left in by the previous user. Your goal at this point is to return to the Main Menu. Press "Esc" until it appears. Here, for example and for future reference, is the Main Menu.

```
- Main    Menu -
1.Data Capture
2.Position
3.GPS Status

4.Navigation
5.Date & Time
6.Configuration
7.Data Transfer
```

---

[3] The designation "{__}" indicates that some action is required on your part. A sentence *in italics* following {__} indicates a general activity that is to occur and that is explained by following statements. Perhaps you will want to check off each activity as it is performed.

{__} Press and hold the on-off button. The receiver will not immediately go off. Rather, a screen will appear that tells you that "OFF" is impending in 5, 4, 3, etc. seconds. This "delayed off" serves a purpose. In case you were recording data and accidentally switched the receiver off, you get a chance to recover from this error by simply releasing the button. (If, while the screen is counting down, you press "CMD", "off" takes effect immediately. Practice that now.)

{__} Turn the receiver on again. Darken the room or move to a relatively dark place. Hold the on-off switch down for one second. Note that this causes a screen "backlight" to come on. Repeat the process to turn the light off.

Be aware that, if you are using battery power, the backlight reduces the amount of time you can use the unit without recharging the battery pack.

{__} With the receiver on, press and hold the on-off key and the "up" key together. Note the change in contrast on the screen. Release both keys. Now try the same with the "down" key. (This process will also turn on the backlight, so be careful not to leave the backlight on and drain the battery.) Turn the receiver off.

{__} Now cycle through this on-off procedure a couple of times, employing the backlight also, until you get comfortable with it (and have noted down the information on the initial screens).

## The Sources of Power and Other Gadgets

{__} While you are exploring the receiver, get acquainted with the other paraphernalia that comes with it. You have already met the rechargeable battery pack. Fully charged, it should supply more than eight hours of reception if the backlight is not used. It charges in about the same amount of time as it takes to run down.

{__} A second source of electricity for the GeoExplorer receiver is a detachable battery pack that holds four "AA" size batteries. Using disposable alkaline batteries, it powers the receiver for about 2 hours; if rechargeable batteries, or standard flashlight batteries are used, the operating time is less.

When either of the battery packs is almost exhausted, the screen fades and the receiver shuts itself down. Almost no notice is given before the receiver shuts off, so treat the amount of charge in the battery pack conservatively, as you might the amount of gas in your tank on a long trip, with few fueling stations along the way. For information about the battery pack (warnings, how to charge it, its ability to hold a charge while on the shelf, or replacing the "AA" cells) see the receiver operating manual.

A third option for operating the receiver is to plug the power cord into a car, boat, or aircraft DC power outlet. The connector is designed to fit a standard cigarette lighter. The acceptable range of voltage input is broad: from 9 to 32 volts. While this is a good way to power the receiver when such power sources are available, you must be careful not to start or stop the engine of the vehicle while the unit is plugged in, whether or not the receiver is off. Starting an engine, in particular, can induce voltage spikes that can damage the receiver, *even if it is turned off.*

The other accessories you will find with the receiver should be:

- The battery charger, with two red LED indicators. You may leave the battery in the charger indefinitely. The battery is fully charged when the "FAST CHARGE" LED goes out.
- A cable that can connect the receiver to a computer. At the receiver end is a circular eight-pin, male plug. At the other (computer) end is a standard nine-pin RS232 serial port connector.
- Some PC computer software — either GEO-PC or PFINDER — and a printer-port hardware key, to be discussed later.
- Manuals related to the GeoExplorer: Operation Manual, Software User Guide, and a General Reference.

**Understanding the Screens and Controls**

The process of giving commands to the GeoExplorer consists of selecting choices from a "Main Menu" and its sub-menus, and then pressing the CMD key. In other words, you may freely press the arrow keys without changing anything but the display. And you may "escape" from whatever menu you are viewing to a higher-level menu by pressing "Esc". When you press the CMD key you are selecting from a menu. Sometimes this simply brings up new choices. But sometimes it changes the configuration or action of the receiver.

The menu structure of the GeoExplorer is strictly hierarchical. The following figure describes it.

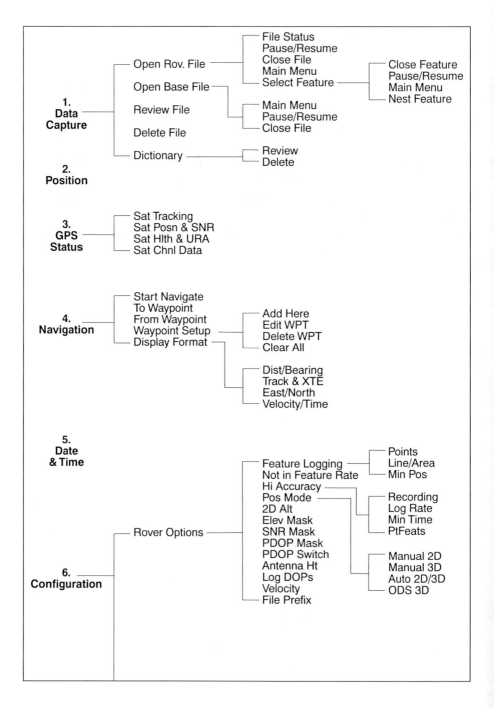

Fig. 1-4 — GeoExplorer menu structure

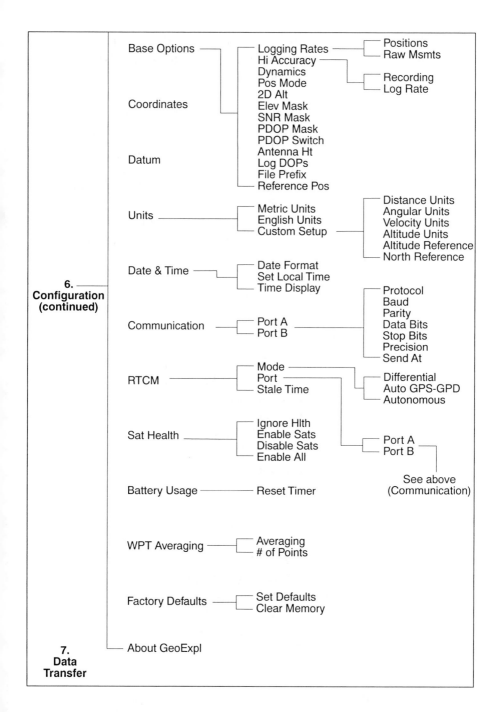

Fig. 1-4 (continued) — GeoExplorer menu structure

To get to a particular menu item from the main menu, you simply select the appropriate menu items that lead to it. We first practice this by returning the receiver to its factory defaults.

{__} Make sure the GeoExplorer is on. Press "Esc" until "Main Menu" appears on the top line. On the third line of the screen you will notice a sequence of dark rectangles — flashing on and off. This is a **highlight** that identifies a **field** which may be either modified or selected by pressing buttons on the GeoExplorer.

{__} Press and hold the "down" key. Note that the main menu "scrolls" quickly, allowing you to find the item you want without repeatedly pressing the button.

{__} Press the "down" key until the highlight appears over menu item "6.Configuration". Press "CMD".

```
        - Configuration -
         1.Rover Options
         2.Base Options
         3.Coordinates

         4.Datum
         5.Units
         6.Date & Time
         7.Communication
         8.RCTM
         9.Sat Health
        10.Battery Usage
        11.WPT Averaging
        12.Fact.Defaults
        13.About GeoExpl
```

**Navigate** through the menu system so you can reset the receiver to the "Factory Default" settings. To do this, press "CMD", then highlight "12.Fact. Defaults". Press "CMD".

{__} You are presented with a screen which will allow you to either "Set Defaults" or "Clear Memory." If you didn't want to do either, you could press "Esc" to return to the previous menu. Instead, use the arrow keys to make sure that highlight is over "Set Defaults." Press "CMD". Resetting the receiver in this way is a fairly serious step, so the GeoExplorer gives you a last chance to back out. Double-check that you are not about to clear memory, then press "CMD". The message should indicate that the receiver has been reset to its defaults. Press "Esc".

{__} Return the unit to the "Main Menu." Highlight and select "5.Date & Time". A screen something like this will appear:

```
DAY mm/dd/yy
 hh:mm:ss UTC
GPS Week: nnn
Local-UTC:+00:00
```

On the top line you see the current day and date (actually, it's the current day and date in Greenwich, England). The world's time standard was previously called Greenwich Mean Time (GMT). The name for it now is Coordinated Universal Time, abbreviated UTC,[4] and it appears on the second line of the screen. UTC is based on a 24-hour clock. The third line shows the "GPS Week." About how many years would you say have elapsed since the official start of the NAVSTAR GPS project? _____. In what year was GPS Week number one? _____ The fourth line is a formula which declares that "Local minus UTC time is zero." Unless you are in England, that is clearly a lie.

---

[4] It is called UTC (rather than CUT) because it is based on international agreement and, in French, the adjective "coordinated" comes last.

{__} To correct the receiver so it shows true local time, and to practice some more with menus, navigate to:

Configuration ~ Date & Time ~ Set Local Time[5]

Note the fourth line of the screen:

Local – UTC: 00:00

with a highlight flashing over the rightmost two digits. Use the "down" key to make the formula read:

Local – UTC: – 05:00

(Note the two negative signs.) This sets the local time to Eastern Standard Time (EST). Reason as follows: EST is five hours earlier than UTC time. For example, when it is 11 a.m. in Greenwich it is 6 a.m. in the eastern U.S. Six minus eleven equals negative five.

Greenwich time does not change when daylight savings time begins. So the difference between EDT (Eastern Daylight Time) and UTC is four hours.

{__} Using the facts above, and knowing the difference between your local time and Eastern Time, set the correct formula on the screen. Press "CMD".

{__} Move to "3.Time Display". Select Local 12 Hours. Press "CMD".

{__} Return to "Date & Time" on the *Main Menu* to verify that they are set correctly. The time shown on this screen should be the correct local time, assuming your receiver has taken data from the satellites recently. If the receiver has been indoors for some time, the time shown may be off by a few seconds. If the receiver's time differs from your local time by a large amount (fifteen minutes or

---

[5] From now on, when you are to navigate to a certain menu item, I will use this "~" notation. For example "**Item1 ~ Item2 ~ Item3**" means: Highlight Item1, press "CMD", highlight Item2, press "CMD", highlight Item 3."

more), you need to rethink the formula you entered on line four under "Configuration".

{__} Write down the information under

"Configuration ~ About GeoExpl."

Compare this information with your notes from the screens that appeared when you turned on the GPS receiver.

## Setting Vital Parameters

Several settings under "Configuration" must be made correctly; if not, data collection may be hampered or nullified.

{__} Go to "Configuration ~ Rover Options". This is a long menu, but many of the items will not apply to your data collection at this time.

```
┌─────────────────────────────┐
│     - Rover Options -       │
│    Feature Logging          │
│      Points                 │
│      Line/Area              │
├─────────────────────────────┤
│     Min Posn                │
│    Not in Feature           │
│      Rate                   │
│    High Accuracy            │
│      Recording              │
│    Log Rate                 │
│      Min Time               │
│      PtFeats                │
│    Dynamics                 │
│    Pos Mode                 │
│    2D Alt                   │
│    Elev Mask                │
│    SNR Mask                 │
│    PDOP Mask                │
│    PDOP Switch              │
│    Antenna Ht               │
│    Log DOPs                 │
│    Velocity                 │
│    File Prefix              │
└─────────────────────────────┘
```

{__} Highlight the field associated with "Dynamics" and press "CMD". Select "Sea" and note the results. Now change the dynamics to "Land" mode. The dynamics selection affects the way the receiver tracks satellites. For example, in "LAND" mode it is expected that the receiver will move more slowly than in "AIR" mode. We provide a more complete description later.

{__} Make sure the "Pos Mode" is set to "Manual 3D". This "position fix mode" should always, repeat, always, be set to "Manual 3D" unless you have enough knowledge and a particular need to set it to something else.

{__} Set the PDOP Mask to "6" by pressing "CMD" when the highlight is on the currently displayed PDOP Mask value. Use the "up" and "down" keys to change the value, and finally press "CMD".

A "**mask**" is a user set value. The receiver compares a given mask with another value that is automatically computed by the receiver. Based on the outcome of the comparison, the receiver uses (or doesn't use) a satellite (or a set of satellites called a "**constellation**") in calculating a position fix. That is, a mask "blinds" the receiver to certain satellites whose signals or positions do not meet the proper criteria for good position finding. We will explain the "PDOP" (Position Dilution Of Precision) term later. For now, just be aware that any PDOP over eight is unacceptable, and six is a figure to use for really precise positioning finding.

{__} Make sure the "Elev Mask" is set to "15". The elevation mask dictates that no satellite below the number of degrees (measured with the horizon considered as 0 and the zenith considered as 90) specified will make a contribution to finding a position. For a roving receiver, as this one will be when you get it outside, a good value is 15°.

{__} Select the "SNR Mask". Each satellite used to compute a position should have a "**Signal-to-Noise Ratio**" (sometimes referred to as signal strength) of four or greater. Verify or set this parameter correctly.

## Preparing to Correlate GPS Data with Map Data

{__} If you are going to take data in the United States, obtain a USGS topographic quadrangle (a topo map, usual scale 1:24,000) of the appropriate general area.

The receiver can display a geographic position in several coordinate systems. Under "Configuration ~ Coordinates" you will find

- Degrees, Minutes, and decimal fractions of minutes (Deg & Min)

- Degrees, Minutes, Seconds, and decimal fractions of seconds (Deg, Min & Sec)
- Ordinance Survey of Great Britain (OSGB)
- Universal Transverse Mercator (UTM)
- Earth Centered, Earth Fixed (ECEF)
- Trimble Grid

{__} Choose "Deg & Min".

{__} Under "Configuration ~ Datum" you will find a list with a large number of choices, stored in alphabetical order. You can scroll through the list by holding down either the "up" key or the "down" key. Choose Bahamas[6] (NAD-27) with the "CMD" key.

{__} Then choose WGS-84 (World), which is the fundamental GPS datum. **WGS-84** identifies the World Geodetic System developed in 1984.

{__} Finally, note the datum of the map you are using. The datum is usually found in the lower left-hand corner (e.g., NAD-27, the North American Datum of 1927). Under "Configuration ~ Datum" set the GPS receiver to operate in this datum, shown as "N-Am. 1927 Conus".

{__} Under "Configuration ~ Units ~ Custom Setup" you will find this menu:

```
- Custom Setup -
1.Distance Units
2.Angular Units
3.Velocity Units

4.Altitude Units
5.Alt Reference
6.NorthReference
```

---

[6] You can dream, can't you?

{__} Under "Distance Units" choose Kilometers. Distance units available on the GeoExplorer include Yards, Meters, Kilometers, Nautical Miles (6080 feet), Miles (statute, 5280 feet), "Internatl Feet" (international feet, where an inch is 0.0254 meters, exactly), and "U.S. Survey Feet" (where a meter is considered to be 39.37 inches, exactly).

{__} Examine your map to determine the appropriate distance units for the upcoming fieldwork. Feel free to change this value later if you should wish visual output in some other units.

{__} Under "Configuration ~ Units ~ Custom Setup ~ Altitude Units" choose Meters or Feet, depending on the map you will be using.

{__} Under "Configuration ~ Units ~ Custom Setup ~ Altitude Reference" choose Geoid (MSL). The "altitude reference" may be set to either MSL (which is elevation above Mean Sea Level) or HAE (which is Height Above the reference Ellipsoid — a theoretical mathematical surface that approximates the surface of the Earth). Since at this time you probably do not know the relationship between the HAE and MSL at your location, you are selecting MSL.

{__} Under "Configuration ~ Units ~ Custom Setup ~ North Reference" choose "North, True" or "South, True" depending on the map you will be using.

## Double checking the configuration

{__} Now do a summary check of the configuration. Under the menu choice "Main ~ Configuration" set the following:

- `Rover Options ~ Dynamics: Land`
- `Rover Options ~ Pos Mode: 3D`
- `Rover Options ~ Elev Mask: 15`
- `Rover Options ~ SNR Mask: 4`
- `Rover Options ~ PDOP Mask: 6`
- `Rover Options ~ Antenna Ht: 1.00 (or zero,`
  `                    or your height in meters)`
- `Rover Options ~ Log DOPs: Off`
- `Rover Options ~ Velocity: Off`
- `Rover Options ~ File Prefix: (do not change)`
- `Rover Options ~ Not in Feature Rate: 3s`
- `Coordinates ~ Deg & Minutes`
- `Datum: (set to your map)`
- `Units ~ Custom ~ Distance: (set to your map)`
- `Units ~ Custom ~ Angular: Degrees`
- `Units ~ Custom ~ Velocity: Miles per Hour`
- `Units ~ Custom ~ Altitude Units: Meters`
- `Units ~ Custom ~ Altitude Reference: Geoid (MSL)`
- `Date & Time ~ Set Local Time: (adjust to local time)`
- `Date & Time ~ Time Disp: Local 12 Hours`
- `Battery Usage: (reset if battery newly charged)`

Use this checklist each time you take the GeoExplorer into the field.

## Final Inside Activity

You are almost ready to take the GeoExplorer into the field. One thing remains to be done. While you are still inside, read through the directions for PROJECT 1B below **completely** to prepare yourself for the fieldwork. Develop a feel for the sort of data you will be collecting. Practice changing from screen to screen. Outside, with the wind blowing and the traffic roaring, is no time to discover that you don't have a solid surface to write on or that you don't know just what it is you are supposed to be doing. A little preparation now will pay big dividends later.

{__} Reading completed on PROJECT 1B.

## PROJECT 1B

## Now Outside

This is an exercise best done with two people. You will take the map, your notebook, and the GPS receiver outside to make observations. You will not yet place the data you collect into a computer file but you will learn a lot about the factors affecting data collection. (If you are not sure that the settings on the receiver are those you put in during Project 1A, verify them against those in the previous section: "Double Checking the Configuration.")

{__} As you leave the classroom or laboratory to travel to the site for data collection, turn the unit on. If you carry the receiver exposed to the sky, it will begin to "acquire" satellites. It is not important which menu appears on the display; whenever the receiver is on, it "looks" for satellites and calculates positions if it can.

{__} Move to a spot outdoors, well away from buildings and heavy tree canopy. If it is reasonably level and not shrouded by nearby hills or mountains, so much the better. And if you can locate the antenna over a geodetic monument, for which you can find the official latitude and longitude, super.

{__} Look at the map to locate your approximate position.

{__} Place the antenna over the spot for which the coordinates are to be determined. The antenna in the receiver is just below the image of the sextant embossed in the plastic of the top of the receiver. Hold the unit as far in front of you as is comfortable, with the top part close to horizontal, tilted only enough so you can read the screen.

Actually, no position will be comfortable after a few minutes; you will want to pass the unit to your partner so you can drop your arm and let the blood drain back into your fingers. An alternative is to put the unit on the ground and crouch or sit down so you can read the screen. This is less fun in winter, or when there is poison ivy about. We never claimed fieldwork was easy. You might bring a table or tripod with you, or find a fence post. Be careful: the power cord makes it easy to bounce the receiver off the ground. It's a tough unit but it is also expensive; do you really want to test it?

{__} As important as finding a good site is keeping your head and body out of the way, i.e., don't block the signal from a satellite to the receiver. You are opaque, as far as the high-frequency, short-length GPS waves are concerned. Remember, the receiver is looking for satellites as low as 15 degrees above the horizon. It's easy to forget this and obstruct the antenna, causing the receiver to lose its lock on a satellite.

**Tracking Satellites**

{__} Starting with the Main Menu, navigate to the "GPS Status" menu and press "CMD". The following will appear:

```
┌─────────────────────────┐
│  - GPS Status -         │
│  1.Sat Tracking         │
│  2.Sat Posn & SNR       │
│  3.Sat Hlth & URA       │
├─────────────────────────┤
│  4.Sat Chnl Data        │
└─────────────────────────┘
```

Highlight "Satellite Tracking" and press "CMD". The "Sat Tracking" screen will appear, listing some two-digit numbers. These are the designations, called **PRN numbers,**[7] that your GPS receiver uses to identify the satellites. The numbers lie between 1 and 32, inclusive.

The numbers which appear now are those which the receiver might be able to pick up, based on your position and time. They are usually those which are above the horizon and the specified elevation mask angle. The receiver determines which satellites are available by formulas built into its computer and by an **almanac** transmitted by each satellite which describes the general location of all the satellites.

Since you are outside, presumably the receiver is locked onto some satellites. The number of little shaded boxes in the lower left-hand corner of the screen indicates how many. The receiver needs to be receiving at least four satellites before location fixes are computed.[8] It may receive signals from as many as eight.

One more bit of information may be learned from this screen. Small arrows — up to four of them — may be seen pointing to satellite numbers. These are the satellites which the receiver is using (or attempting to use) to calculate its position.

---

[7] "PRN" stands for "Pseudo Random Noise" (no doubt just the designation you expected for the satellites). I will explain later.

[8] If the receiver is in 2D mode, only three satellites are required for a location fix. But unless you have entered the precise altitude of the antenna, the locations calculated by the receiver will be wrong. Again, don't use 2D mode unless you know what you are doing.

To summarize the screen: If a satellite's number appears on the screen, the satellite is physically in the space above the user, at an angle greater than the setting of the elevation mask, where it might be used for positioning finding. If an arrow appears next to the number, it is being used for position finding. The number of boxes indicates the number of satellites that the receiver is electronically locked onto.

A few minutes may elapse before the unit locks onto enough satellites to begin giving location fixes. If more than 10 minutes go by with no location fix, change the PDOP to eight (8).

{__} Once the GeoExplorer is tracking four or more satellites, select "Position" from the Main Menu and write down the latitude, longitude, and altitude. When locked onto four or more satellites, the receiver computes the position of the antenna about three times every two seconds. (If the word "OLD" appears on the screen it indicates that the value presented is one that was collected in the past — perhaps the immediate past — and that the receiver is not calculating new positions. Make certain that there are no obstructions blocking the signals.)

{__} Note the time. Plan to write down a new position reading in your notebook every minute, approximately on the minute, for the next quarter of an hour.

{__} In between writing position fixes in your notebook you should record some other information. Move back to the "Sat Tracking" screen. Note down the numbers of the satellites which appear there. Circle the numbers of the satellites the receiver is using to compute positions. Also note how many satellites the unit is receiving signals from. Write down the value identified as "PDOP".

{__} Now it is probably about time to go back to the "Position" screen to write down the next set of position coordinates. They should be close to, but not exactly the same as, those you wrote down a minute ago. The screen should not say "Old Position". If it does, you probably got your head in the way of a satellite signal.

{__} Now go to the "Sat Posn & SNR" screen (it's under "GPS Status"). You will see several horizontal lines of information — one for each satellite being tracked. One item of information displayed for each satellite is **"Elv"** — an abbreviation for **Elevation:** If you could stand and point a straight arm directly toward the satellite, the elevation would be the angle, in degrees, that your arm made with the Earth, assuming the surface is level where you are standing. Zero degrees would represent a satellite at the horizon; ninety degrees would represent a satellite directly overhead.

{__} Whoops. Time to write down another lat-lon-alt position.

{__} Return to the "Sat Posn & SNR" screen. The column after "Elv" is identified as **"Az"** which stands for **Azimuth.** The Azimuth specifies the angle between due north and the satellite: Point your arm toward the north, then rotate your body clockwise until your arm is pointed at the satellite. The number of degrees your body rotated is the azimuth.

{__} Write down another position fix.

The last column on the "Sat Posn & SNR" screen is the "signal to noise ratio" (recall that it is an indication of the strength of the signal from the satellite). Acceptable values are greater than or equal to four — which is the value you set as the SNR mask. Values may range up to 35 or so.

{__} For each satellite being tracked, record its elevation, azimuth, and signal strength.

{__} Put your hand over the antenna (it is directly under the little sextant embossed in the plastic above the screen) and watch the signal strength drop.

{__} Determine where one or two satellites are in the sky, relative to your position. Try to interpose your body between the unit and a satellite to see if you can make the signal strength drop for a single satellite. In the middle latitudes in the U.S. there will generally be fewer satellites to your north than south.

{__} After recording another fix, move to the "Sat Hlth & URA" screen. This displays the "health" of the satellite, as determined by information broadcast by the satellite itself, and the "User Range Accuracy" — a numerical indication of the accuracy one might expect when using this satellite to compute a position fix.

Satellite health will be indicated by "OK", by "U" for unhealthy, or by "n/a" for "not available", indicating that no signal is being received.

"URA" may have values ranging from one to 1024. Values greater than 16 indicate that the DoD is corrupting the signal for the particular satellite and that any single point calculated by using that satellite could be in error by approximately 100 meters. The units of the URA number are meters, but since a given position is found using a combination of several satellites, the URA value of any particular one is of limited usefulness in estimating error.

{__} Finish recording the fifteen position fixes. Is the unit still tracking the same satellites? Is it using the same constellation of satellites to compute fixes? If not, write down the new information.

## Set Your Watch

The GPS receiver's clock has been reset by the exposure to the satellites. It now has a very accurate idea of the time. So you may set your watch by it and be correct to the nearest second.[9] With the averaging of the coordinates and the setting of your watch, you have used GPS to position yourself in four-dimensional (4D) space.

---

[9] The UTC time known by the GPS receiver is correct to better than a billionth of a second, but the display is not nearly that accurate.

## Did the Earth Move?

{__} Go to the "Navigation" item on the Main Menu. When you select it you will find five menu items, as shown below:

```
1.Start Navigate
2.To Waypoint
3.From Waypoint
4.Waypoint Setup

5.Display Format
```

{__} Select "Display Format". Select "Velocity/Time". The display returns to the "Navigation" menu. Select "Start Navigate". Scroll the screen until "Vel:" (for Velocity) appears. A number, representing kilometers per hour (kph), is displayed. The number will be between zero and around five.

Any number greater than zero indicates that the antenna is moving at some number of kilometers per hour with respect to the Earth. That's odd. You see that the antenna is virtually motionless. Why should the receiver be recording movement? Because, in addition to the errors that occur naturally, the U.S. DoD may be distorting the signal. They would rather that, for example, an enemy soldier with a mortar not know exactly where he is. The term for this deliberate perturbation of the signal is "**selective availability**". Later, you will learn how to get accurate measurements to put into a GIS, despite selective availability.

{__} Now begin to walk with the unit held out in front of you. Call out the velocity readings to your partner. He or she should "mentally" average your readings and record some values. A comfortable walking speed is about five kph (three miles per hour). Is that what the unit indicates?

{__} Continue to walk. Escape from the current screen and select "Display Format" again. This time pick "Dist ~ Bearing". Select

"Start Navigate" again. Scroll the screen until "Heading:" appears. The number indicates your direction of travel, relative to True North ("Tn"), in degrees. Again, call the readings out to your partner. Do they tend to average to the approximate direction you are walking?

{__} Walk back to the original location where you recorded the position fixes. After a minute or so, shut the unit down and return to your base. Be certain the screen is blank, so that the unit is not collecting new position fixes. Because later we will want to use the last position recorded, *do not turn the unit on again* until you are inside a building.

## PROJECT 1C

### Back Inside

Your session in the field may have raised as many questions as it answered. We will look at the answers to those questions in later chapters. First, let's verify that GPS really works. (Someone is telling you that you can find your position on Earth to within a few feet from four objects in space, 12,000 or more miles away, batting along at 2.4 miles per second. Would you believe them without checking? I wouldn't.)

{__} Using a calculator, obtain the average of each of the fifteen latitudes, fifteen longitudes and fifteen altitudes you recorded. Plot the point on your topo map. Does the point represent where you were?

The altitude indicated by the GPS receiver is likely to be very different from that shown by the map. The horizontal accuracy of a single point is usually within 100 meters, or 300 feet. Vertical accuracy is about half that good. So your altitude fix on the last point your receiver recorded could be off by 600 feet. The average of the 15 altitudes should be somewhat better.

## Two Altitude Referencing Systems

For all but the last few years, most people measured altitude from the average level of the oceans. The two primary methods of taking measurements were vertical length measurement from a beach (inconvenient if no ocean were nearby) and measurement of air pressure. Of course, air pressure is variable from hour to hour so there are complications using this method as well.

Satellites are kept in their orbits by gravity. Gravity can be considered a force between the satellite and the center of mass of the Earth. For this reason, a new definition of altitude has been developed, using not sea level as the zero but a "gravitational surface" called the "reference ellipsoid." The geometrical object chosen to approximate the Earth is an ellipsoid, rather than a sphere, because the Earth more resembles a ball which has been compressed slightly at the poles, so that it bulges at the equator. The diameter through the poles is some 43 kilometers less than a diameter across the plane of the equator. The reference ellipsoid approximates mean sea level, but is slightly different from it almost everywhere. Since the satellites are slaves to gravity, the GeoExplorer "thinks" in terms of height above the reference ellipsoid, but has in it formulas and tables which allow it to display results in MSL.

{__} Now determine the approximate difference between Mean Sea Level and the Height Above Ellipsoid in your area. This information has been coded in your receiver. First, write down the elevation of the last point (the OLD position) from the "Position" screen. This will be a height above sea level in feet or meters, depending on how you set the units. Now under "Alt Reference" select "Ellipsoid (HAE)". Return to the "Position" screen. The new number for elevation shown there is the height of the OLD position above the reference ellipsoid. From these two numbers you can calculate the difference in height between the reference ellipsoid and mean sea level. What is it and in what units? _____ Which is higher? _____

## The Datum Makes a Difference

It is absolutely vital, when integrating GPS data with GIS data,
that your datasets match with respect to geodetic datum, geodetic
datum,[10] coordinate system, units, and projection. You will soon
prove to yourself how important this is by filling out the Latitude
and Longitude Computation Tables below.

Some relevant information about determining shifts in position
designation from one datum to another, and from one coordinate
system to another, are:

- One degree of latitude corresponds to approximately
  111 kilometers (km); therefore one minute of latitude
  corresponds to approximately 1845 meters.
- The length of a minute of longitude along a parallel
  depends upon the latitude of that parallel. The length
  varies from approximately 1855 meters at the equator
  to zero meters at the poles. So some computation is
  needed: at the latitude shown by your receiver for the
  OLD position (_____), one minute of
  longitude corresponds to approximately 1855 meters
  *multiplied by the cosine of the latitude* (e.g., the cosine
  of the latitude 30° is approximately 0.866) which, at
  the position of your fix, is _____. Thus, a
  minute of longitude in your area implies
  _____ meters.
- On the Trimble display, position information in UTM
  is given with Easting (the x-coordinate) first, and
  Northing (the y-coordinate) second. Note that this is
  opposite the "latitude first" convention of "Deg &
  Min".
- In the UTM coordinate system, in a given zone, a
  *greater* number of meters indicates a more easterly
  position in longitude, or a more northerly position in
  latitude. That is, "x" and "y" increase "to the right"

---

[10] **"Datum"** is the master geodetic referencing system used for a particular
project or map.

and "up," respectively, in accordance with standard Cartesian convention.

{__} Fill out the Latitude and Longitude Computation Tables below. To do this, you will have to take eight readings from the receiver. The receiver kept the coordinates of the last data point you took in the field. You can view the latitude and longitude of that point using whatever datum you want and in whatever coordinate system you want. You will take the eight readings by alternating between the "Position" screen and the configurations that allow you to specify datum and coordinate system. When you are finished taking the readings and doing the calculations, come back and answer the question below:

{__} Does the difference in meters from WGS-84 to NAD-27 using the UTM coordinate system correspond to the difference in meters you calculated based on degrees and minutes? _____ What conclusions can you draw from your observations and calculations?

## LATITUDE COMPUTATION ON "OLD" POSITION

### Degrees & Minutes

NAD-27            _____ (N-Am. 1927)

WGS-84           _____

Difference        _____

In this area, the representation in "Deg & Min" of the point in WGS-84 (alias NAD-83) is _____ minutes (north or south) _____ of the representation of that point in NAD-27. This corresponds to _____ meters.

~~~~~~~~~~~~~~~~~~~~~~~~~~~~~~~~~~~~~~~~~~~~~~~~

### UTM (in meters)

NAD-27            _____

WGS-84           _____

Difference        _____

In this area (which is UTM Zone: Number ___, Letter ____), the representation of the point in UTM coordinates based on WGS-84 is _____ meters (north or south) _____ of the representation of that point in UTM NAD-27.

## LONGITUDE COMPUTATION ON "OLD" POSITION

### Degrees & Minutes

NAD-27                  _____

WGS-84                  _____

Difference              _____

In this area, the representation in "Deg & Min" of the point in WGS-84 (NAD-83) is _____ minutes (east or west) _____ of the representation of that point in NAD-27. This corresponds to _____ meters.

~~~~~~~~~~~~~~~~~~~~~~~~~~~~~~~~~~~~~~~~~~~~~~~

### UTM (in meters)

NAD-27                  _____

WGS-84                  _____

Difference              _____

In this area (which is UTM Zone: Number \_\_\_, Letter \_\_\_), the representation of the point in UTM coordinates based on WGS-84 is _____ meters (east or west) _____ of the representation of that point in UTM NAD-27.

## GPS Equipment Check-out Form

We (please print names, telephone numbers, e-mail addresses)

_____,_____,_____

_____,_____,_____

wish to check out the GPS receiver (check one)

( )GeoExplorer  ( )Pathfinder Basic  ( )_____

on _____ ____/____/____ at _____ at _____.
   (day)     (date)      (time)     (place)

We understand that this equipment is valuable, and we will use our best effort to guard it against damage, loss, or theft.

We intend to take data at _____
which is at approximately latitude (DMS) _____
and longitude (DMS) _____.

We will return said equipment

on _____ ____/____/____ at _____ at _____.
   (day)     (date)      (time)     (place)

_____        _____
     (signed)                   (signed)

In case of difficulty, contact _____.
Leave a message on the recorder if there is no answer.

# Part 2

# Automated Data Collection

IN WHICH *you learn the basic theoretical framework of GPS position finding, and practice using a GPS receiver to collect computer-readable data.*

# Automated Data Collection

## OVERVIEW

### How'd They Do That?

By now you have determined that GPS really works. That little box can actually tell you where you are. How?!

The fundamentals of the system are not hard to comprehend. But misconceptions abound, and it is amazing how many people don't understand the principles. In a few minutes, assuming you keep reading, you will not be among them.

We look first at a two-dimensional analogy. You are the captain of an ocean-going ship off the western coast of some body of land. You wish to know your position. You have aboard:

- an accurate timepiece;
- the ability to pick up distant sound signals (a megaphone with the narrow end at your ear, perhaps?);
- a map showing the coast and the locations of any soundhouses (a soundhouse is like a lighthouse, but it emits noise instead); and

- the knowledge that sound travels about 750 miles per hour, which is about 20 kilometers (km) per minute, or 1/3 km per second.

Suppose:

(1) there is a soundhouse located at "S1" on the diagram below.

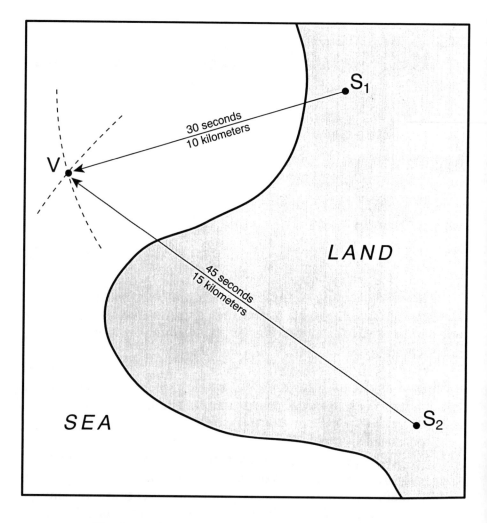

Fig. 2-1 — Measuring distance by measuring time

(2) Each minute, exactly on the minute, the soundhouse horn emits a blast.

With these elements, you can determine the distance of your vessel, "V", from the soundhouse. To do so, you note when your clock marks an exact minute. Then you listen for the sound signal. When it comes, you again note the second hand of the clock. You then may calculate "d" from

$$d = s/3$$

where "d" is the distance in km from the soundhouse and "s" is the number of seconds it took for the sound to reach you.

Suppose it took 30 seconds for the signal to arrive. You would know your ship was 10 km from the soundhouse.

In geometric terms, you know only that your ship is located somewhere on the surface of a sphere that has a radius of 10 km. You can reduce this uncertainty considerably since your ship is floating on the ocean, so you know your altitude is fairly close to mean sea level. Thus you could consider yourself to be on a circle with a radius of 10 km.

That, however, does not pinpoint your location. And, by looking at the diagram, you can see that, if you are moving, you might be in a lot of trouble, since contact with the ground is not recommended for ships. How could you determine your position more exactly? By finding your distance from a second soundhouse.

Suppose you use the same technique as above, listening to soundhouse "2". You find it is 15 km away. Then the question is: where is (are) the point(s) that are 10 km away from "1" and 15 km away from "2"?

If you solve this problem graphically, by drawing two circles, you find they intersect in two places. One of these locations you can pretty well eliminate by noting that your ship is not sitting on a prairie or in a forest. Based on the measurement from the second soundhouse, you know your position as accurately as your measuring devices and map will let you know it.

It is obvious that you can find your position by drawing circles. You can also find your position by purely mathematical means. (You would determine the coordinates of the intersection of the two circles from their formulas by solving two "simultaneous" equations. Then you would consider lengths of the sides of the triangle formed by the two soundhouses and the ship. This process is sometimes called "trilateration." (A triangle is a "trilateral," as a four-sided figure is a "quadrilateral.")

As with the soundhouses, distances — determined by the measurement of time — form the foundation of GPS. (Such distances are referred to as **ranges**.) Recall the basis for the last three letters of the acronym NAVSTAR: Time And Ranging.

## How it works: measuring distance by measuring time

How do the concepts illustrated above allow us to know our position on or near the earth's surface? The length of that answer can vary from fairly short and simple to far more complicated than you are (or I am) interested in. First let's look at the most basic ideas by seeing how the "ship" example differs from NAVSTAR.

- NAVSTAR gives us 3D locations: Unless we are in fact on the sea, in which case we know our altitude, the problem of finding our location is three-dimensional. GPS can provide our position on or above the earth's surface. ("Below" is tricky because of the radio wave line-of-sight requirement.) But the method translates from two to three dimensions beautifully.
- The "soundhouses" are satellites: Rather than being situated in concrete on a coast, the device that emits the signal is a satellite, zipping along in space, in a quasi-circular orbit, at more than two miles per second. It is important to note, however, that at any given instant the satellites (space vehicles, or SV's) are each at one particular location.
- NAVSTAR uses radio waves instead of sound: The waves that are used to measure the distance are electromagnetic radiation (EM). They move faster than sound — a lot faster. Regardless of frequency, in a vacuum EM moves at about 299,792.5 km/sec, which is roughly 186,282 statute miles per hour.

Let's now look at the configuration of GPS with a drawing of true but extremely small scale, starting with two dimensions. Suppose we represent the earth not as a sphere but as a disk (like a coin), with a radius of approximately 1 unit. (One unit represents about 4000 statute miles.) We are living on the edge (pun intended) so we are interested in points on, or just outside (e.g., airplanes), the edge of the coin. We indicate one such point by "x" on the diagram below. We want to find out where "x" is.

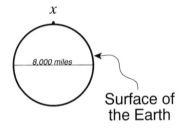

Fig. 2-2 — Earth and a point to be found

Suppose now that we have two points on the paper called "a" and "b". They represent two of the NAVSTAR satellites, such that we can draw straight lines from "a" to "x" and from "b" to "x". We require that the lines not pass through our "earth." These points "a" and "b" are each about four units from the center of the coin (thus three units from the edge of the coin).

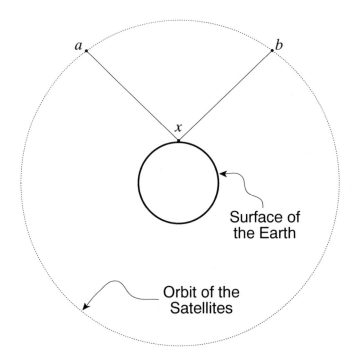

Fig. 2-3 — Earth and satellites

Measure the distance, or length, from "a" to "x" (call it "$L_a$") and from "b" to "x" (call it "$L_b$"). The lines "$L_a$" and "$L_b$" represent the unobstructed lines of sight from the satellites to our receiver antenna. If we know the positions of "a" and "b" and the lengths of "$L_a$" and "$L_b$" we can calculate the position of "x" through the mathematical process of trilateration.

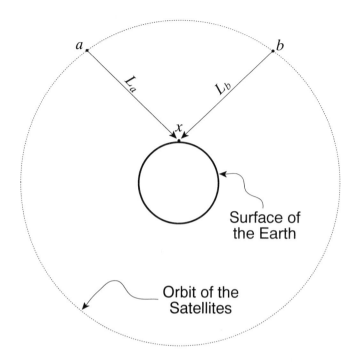

Fig. 2-4 — Distances from "x" to the satellites

It is intuitively obvious that we can locate any "x", given knowledge of the positions of "a" and "b", and of the lengths "$L_a$" and "$L_b$": "x" must lie on a circle centered at "a" with radius "$L_a$" *and* also must lie on another circle centered at "b" with radius "$L_b$".

To formalize a bit what we learned from the example with the ship and the soundhouses, two such circles either (1) do not intersect, (2) touch at a single point, (3) touch everywhere, or (4) intersect at two points. Possibility (1) is clearly out, because we have constructed our diagram so that the circles intersect at "x". Possibility (2) seems unlikely, given the geometry of the situation, and given that the line segments connecting the points with x must not pass through the disk. Possibility (3) is out: the circles would have to have the same centers and radii. We are left then with possibility (4): there are two points that are at distance "$L_a$" from "a" and "$L_b$" from "b". One of these is "x". The other is a long way from the disk, is therefore uninteresting (we'll call it point "u"), and is certainly not "x".

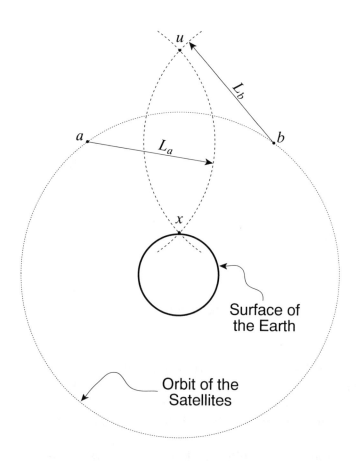

Fig. 2-5 — Finding "x" by the intersection of circles

The points "a" and "b", of course, represent two of the GPS satellites *at a precise instant of time*, and point "x" represents the position of the GPS receiver antenna at that same precise moment.

This is fine for two dimensions. What happens when we move to three? From a conceptual point of view, the coin becomes a sphere, "x" resides on the sphere (or very slightly outside it), and we need to add a point (satellite) "c", somewhere else in space (not in the plane formed by "a", "b", and "x") in order to be able to locate the position of "x". The problem we solve here is finding the intersection of three spheres (instead of two circles).

While the process of finding this intersection is more difficult mathematically, and harder to visualize, it turns out, again, that there are only two points, just one of which might reasonably be "x".

You may deduce this as follows. Consider initially the intersection of the surfaces of two spheres. First, forget about the special case possibility that they don't intersect at all (their centers are separated by more than the sum of their radii, or one sphere completely contains the other) and the case in which they touch everywhere (same centers, same radii). Now they must either touch at a single point (unlikely) or intersect in a circle. So our problem of visualization is reduced to finding the intersection of this circle and the surface of the third sphere. If you again discard some special, inapplicable cases, you see that the intersection can be only two points — only one of which can be "x".

In theory, one should need only three satellites to get a good, three-dimensional (3D) fix. You may recall that when you took the receiver into the field, however, you needed four satellites before the unit would calculate a position. Why? Briefly, the reason has to do with the fact that the clock in a receiver is not nearly as good as the four $50,000 clocks in each satellite, so the receiver must depend on the satellite clocks to set itself correctly. So, in a sense, the fourth satellite sets the receiver's clock. In actuality, however, all four satellites contribute to finding a point in four-dimensional space.

You now know the theoretical basis for GPS. It is not unlikely that you have several questions. I will defer the answers to a few that we have anticipated so that we can take up an issue that bears more directly on the fieldwork you are about to do.

## Factors Affecting When and How to Collect Data

Neither GPS nor any other method can tell you *exactly* where an object on the earth's surface is. For one thing, an object, no matter how small, cannot be considered to be in exactly one place, if by "place" we mean a zero-dimensional point specified by coordinates. Any object occupies an infinite number of zero-dimensional points. Further, there are always errors in any measuring system or device.

Now that we have thrown out the idea of "exact" location, the issue becomes: what kind of approximation are you willing to accept (and pay for)? There are two philosophies you might consider:

- good enough
- the best that is reasonable

You might use "good enough" when you know for certain what "good enough" is. For example, if you are bringing a ship into a harbor, "good enough" might be locational values guaranteed to keep your ship out of contact with the harbor bottom, or between buoys.

You might, on the other hand, use "the best that is reasonable" when you don't really know what "good enough" is. For example, if you are building a database of city block outlines, you might not be able to foretell the uses to which it might be put. Your immediate needs might suggest that one level of accuracy would be appropriate, but several months later you (or someone else) might want a higher level of accuracy for another use. So it might be worth expending the extra resources to collect data at the highest level of accuracy that your budget and the state of the art will allow.

The major factors that relate to the accuracy of GPS measurements are:

- Satellite clock errors
- Ephemeris errors (satellite position errors)
- Receiver errors
- Ionosphere errors (upper atmosphere errors)
- Troposphere errors (lower atmosphere errors)
- Multipath errors (errors from bounced signals)
- "Selective Availability" (SA) errors (signal transmission perturbations created by the NAVSTAR system managers (i.e., deliberately introduced errors))

There are several fifty-cent words above which haven't been defined. We will defer discussion of most of the sources of error until later, primarily because, at the moment, there is little or nothing you can do about them besides knowing of their existence and understanding how they affect your results. (In the long run you can do a lot about these errors — by spending more money for GPS receiving equipment.) But we will consider an issue called "Dilution of Precision" (DOP) because:

- High (poor) DOP values can magnify the other errors,
- DOP values can be monitored during data collection, and
- DOP values, which can be predicted at a given location, can perhaps be reduced by selecting appropriate times to collect data.

## Position Accuracy and DOP

Prior to the end of 1993, GPS had less than the full compliment of 24 satellites operating. In earlier years, there were periods during the day when there were not enough satellites in view from a particular point on the ground to provide a position fix.

Now a data collector can almost always "see" enough satellites to get a position fix. But the quality (accuracy) of that fix is dependent on a number of factors, including,

- the number of satellites in view, and
- their geometry, or arrangement, in the sky.

In general, the more satellites in view, the better the accuracy of the calculated position. The receivers you are using, and many others, use the "best" set of four satellites to calculate a given point — where "best" is based on a DOP value. With more satellites in view, there are more combinations of four to be considered in the contest for "best." Data collection with fewer than five satellites in view is pretty iffy and should be avoided when possible. I'll say more later on how you can know ahead of time if enough satellites are available and what the DOP values will be. It is almost always possible to see four (with 24 up and broadcasting properly — "healthy," as they say) and you might occasionally see more than twice that number.

The concept of DOP involves the positions of the satellites in the sky at the time a given position on the ground is sought. To see why satellite geometry makes a difference, look at the two-dimensional case again. Suppose first there are two satellites ("a" and "b") that are being used to calculate a position "x": If we know the distances $L_a$ and $L_b$ exactly, we can exactly find the point "x".

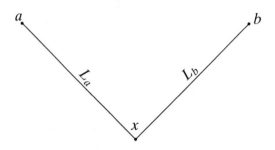

Fig. 2-6 — Satellite positions relative to "x"

But we don't know the $L_a$ or $L_b$ exactly, because of the error sources listed above. For illustration, suppose that we have an error distance, delta, that must be added and subtracted from each of $L_a$ and Lb. That is, for each distance there is a range of uncertainty in the distance that amounts to two times delta. Illustrated graphically this becomes:

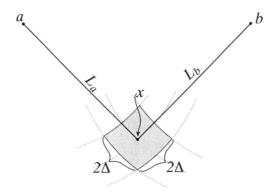

Fig. 2-7 — Area in which "x" might reside
given satellite positions "a" and "b"

The cross-hatched area indicates the surface that contains the actual location sought.

Now suppose there are two satellites ("c" and "d") which are being used to calculate a position "y". These satellites are further apart than "a" and "b", so that the angle their lines make at the receiver are more obtuse than the almost-90 degree angle from "a" and "b".

If we know the distances $L_c$ and $L_d$ exactly, we can exactly find the point "y". But again we do not know the lengths exactly — we use the same difference of two times delta.

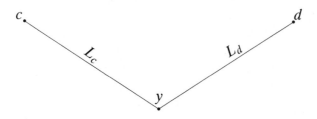

Fig. 2-8 — Satellite positions relative to "y"

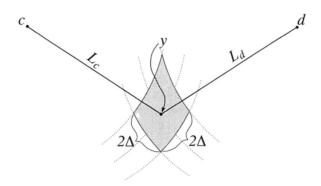

Fig. 2-9 — Larger area of uncertainty, due to satellite positions

The cross-hatched area shows the range of positions the receiver might indicate for fix "y". As you can see, the distance between the true position and the position that could be reported by the receiver for the second case is considerably larger than for the first case.

The goal here is to illustrate that "satellite geometry" can make a big difference in the quality of the position you calculate. In actual GPS measurements, of course, it is a volume rather than an area that surrounds the point being sought, but the same general principles apply.

## So, Actually, What is DOP?

**DOP** — sometimes referred to as **GDOP ('Geometric' Dilution of Precision)** — is a number which is a measure of the quality you might expect from a position measurement of the GPS system *based solely on the geometric arrangement of the satellites and the receiver being used for the measurement.* When a DOP value is unity ("1") the accuracy of the fix is as good as it can be considering all the sources of error. This might occur when one satellite is directly overhead and the three others are close to the horizon, spaced in a circle, separated from each other by 120 degrees. Another way of looking at it is that DOP is a measure of the extent to which satellite geometry exacerbates the other errors that may occur in the measurement.

The overall DOP number is made up of several "sub-DOPs":

- HDOP (Horizontal DOP) is a combination of NDOP (North DOP) and EDOP (East DOP),
- VDOP is Vertical DOP,
- PDOP (Position DOP) is a combination of HDOP and VDOP,
- TDOP is Time DOP, and
- GDOP ("Geometric" DOP) is a combination of PDOP and TDOP.

You may recall that, when you set up the GPS receiver in Part 1 you set the maximum allowable PDOP to "06". (Other than the 2D mask, which you probably won't use, this is the only maximum DOP setting you make.) PDOP is the most important single DOP to consider. The recommended PDOP values are: 1 to 4, great. 5 to 6, okay. As you saw, on one screen you set the PDOP value; on another screen, at the time of data collection, you got a report on the value of PDOP of the constellation the receiver was using. If the PDOP was too high, no positions were calculated.

# STEP-BY-STEP

## PROJECT 2A

### Inside: Planning the GPS Data Collection Session

During the last step-by-step project you visually read and manually recorded positions calculated by the GPS receiver. In this session, using the same geographic location as last time, you will use the memory capacity of the receiver to store the readings in machine-readable (i.e., computer-readable) format. The data will be collected into sets, called *files*.

For the projects in this Part you will need your notebook, the receiver, and the external antenna, if you have it. Set aside a section of the notebook to record information about the files you will collect. On a form such as the one found at the end of this Part, you should plan to manually record:

- the date and time,
- a general description of the location or path that is the subject of your data recording,
- the starting time and the ending time of the data collection,
- the filename,
- the amount of memory available in the datalogger before data collection, and then, when finished collecting a file, the actual size of the file,
- the interval between collected data points (e.g., every 10 seconds, or every 50 meters),
- the values of PDOP,
- the number of satellites being tracked,
- the hours of charge remaining in the battery pack, and the amount of time it is used during this session.

**Setting up the Receiver/Datalogger**[1]

Your goal here will be to ensure that the data you collect during your trip to the field will be worth something when you get back.

The receiver obtains data from the satellites and calculates positions at faster than one per second, but the datalogger usually records point fixes in the microcomputer's memory less frequently. You have some control over how often point fixes are recorded. The datalogger also records the exact time each point fix is taken — a fact whose importance will become apparent later.

{__} Be sure that the battery pack is sufficiently charged to complete the session with an ample reserve.

{__} Ensure that the receiver settings are proper. First check that the settings correspond to those below. Get to the screen "Main ~ Configuration ~ Rover Options" ("Rover" means **roving receiver,** one that may be moved from place to place.) Scroll this menu until "Dynamics" appears.

{__} Set Dynamics: Land

The LAND/SEA/AIR choice relates to a number of factors affecting both display and internal operations of the receiver.

In the LAND mode the receiver is programmed to cope with **"canopy"** such as heavy cloud cover and trees. It is expected that the antenna will be stationary or moving at relatively low velocity (e.g., automobile speed).

In the SEA mode, canopy in the form of clouds is also expected, as is low overall velocity. But since the antenna might be mounted on a mast which could move rapidly from side to side, the receiver is programmed to expect such movement.

In AIR mode the receiver expects the antenna to be moving fairly quickly and that obstructions to the signal will be minimal.

---

[1] For purposes of this discussion, we will speak of the receiver as a device which obtains the fixes from the satellites, and a *datalogger* as a device which records these data into a file. The equipment you are using houses the receiver and datalogger in the same box.

Also, in AIR mode, the display always shows directions in degrees based on magnetic north, rather than true north; pilots navigate on the basis of magnetic north, and runways are numerically designated based on their directions relative to magnetic north.

{__} Set Pos Mode: 3D

{__} Set Elev Mask: 15

{__} Set SNR Mask: 4

{__} Set PDOP Mask: 6

{__} Ignore the PDOP Switch (it is used for 2D settings only)

{__} Set Antenna Ht: 0.00, or to the height you expect the antenna to be above the point whose 3D position you are trying to record.

{__} Set Log DOPs: Off

{__} Set Velocity: Off. This option specifies whether the datalogger is to calculate and record the velocity of the receiver antenna. Velocity data use up memory and are usually not helpful when the antenna is standing still.

{__} For the setting of "File Prefix" ask your instructor or leave it unchanged.

{__} Under the category of "Feature Logging" ignore all settings: "Points", "Line/Area", and "Min Posn". (The GeoExplorer has the capability of recording positions of features (e.g., lampposts), but we won't be using that ability until an exercise in Part 7.)

{__} *Set the "Not in Feature Rate"*: The datalogger usually does not record all the points the receiver calculates. Which points are actually recorded? Basically, you can set it to record points under one of two circumstances:

- when a certain interval of time has elapsed since the last point was recorded, or
- when the receiver senses that the antenna has moved at least a certain distance from the last point recorded.

To experiment with this setting, press "CMD" when the highlight is over the number. On the screen that appears, you may, with the left-right arrow keys, select either of two fields.

The left field may be set to any integer value from 1 to 999, or to "All" or to "Off". (By continually pressing an up-down arrow key, you may scroll through a sequence of numbers; also, individual digits are addressable by use of the left- and right-arrow keys.)

The right field may be set to "seconds", "meters", or "feet".

Since in this exercise you will not be moving the antenna once datalogging starts, set the interval designation units to seconds. Set the number to three (3) and press "CMD".

{__} Double check that the "Not in Feature Rate" is set to three seconds.

{__} Under the category "High Accuracy" ignore all settings: Recording, Log Rate, and so on.

{__} Now you should have scrolled back to "Dynamics", which is where you came in. "Esc". "Esc".

{__} From the Main Menu, select "1.Data Capture".

You now have the following six options:

```
┌─────────────────────────────┐
│   - Data Capture -          │
│   1.Open Rov. File          │
│   2.Open Base File          │
│   3.Review File             │
├─────────────────────────────┤
│   4.Delete File             │
│   5.Rename File             │
│   6.Dictionary              │
└─────────────────────────────┘
```

{__} Select "Open Rov. File".

The data capture screen is transformed; it now looks something like this:

```
┌─────────────────────────────┐
│   R112420A          0       │
│   1.File Status             │
│   2.Pause                   │
│   3.Close File              │
├─────────────────────────────┤
│   4.Main Menu               │
└─────────────────────────────┘
```

The field to the left on the top line is the name of the file the receiver would use if it were outside so it could begin collecting data.[2] (Since the antenna is shielded from satellites, the file will be initiated but no data will be recorded.) The filename consists of eight characters:

**ummddhhi**

---

[2] This is a "default" filename. You may change this name if you wish. I suggest you leave it as it is. If you want a different name, wait until the file is transferred to a computer.

The initial character ("u" for user) of the filename is user-selectable. It is set under "Rover Options". Your instructor may have set it so he or she can keep track of collected student files.

The middle six characters hint at the date and time: the first two digits (mm) are the month, the next two (dd) the day, and the final two (hh) the hour *in UTC time*.

The final character ("i" for index) is one of the sequence "A, B, C, . . . Z, 0, 1 . . . 9 ". Prior to recording the first file of a given hour, the "i" character is set to "A" by the receiver at the beginning of the hour. It is "incremented" to the next character in the sequence each time a file is collected and closed during that hour. This allows several files (up to 36 with the same "u" character) to be collected during a single hour, each with a unique name. If a file is open as the hour changes, *data collection continues under the same filename.*

{__} Check the filename displayed, to be certain it corresponds to the default discussed above, in terms of time and date. Record its name on a piece of scrap paper, since you will want to erase it shortly and you want to be sure you have the right one.

{__} Note the number to the right of the filename. It is the number of data points collected. The number will be zero unless the antenna is outside.

The menu items suggest that you may inquire as to the status of the file, suspend the data collection process, close the file, or return to the main menu.

{__} Scroll along the menu until "File Status" is highlighted. Press "CMD". Scroll this menu until the name of the file appears at the top. The information portrayed should look something like this:

```
┌─────────────────────────────┐
│                             │
│   - File Status -           │
│   File:    R112120A         │
│   Size     0.7519K          │
│   Free     188.82K          │
│                             │
├─────────────────────────────┤
│   #Features        0        │
│   #Positions       0        │
│                             │
└─────────────────────────────┘
```

So you have the filename, the number of fixes (points) in the file, and file size (in kilobytes).[3] Also on this screen the item "Free:" indicates the amount of memory space remaining in the datalogger. Depending on what else is stored in the memory, this number can be anywhere from about 200,000 characters (or **"bytes"**) to none at all.

Another designation is the "Size:" of the file currently being collected. A single 3D fix, with no velocity information recorded, requires 20 to 25 bytes under the best circumstances. Under less than ideal conditions (e.g., interference with the satellite signals — by buildings and trees — which causes the receiver to switch constellations), each point may occupy an average of 100 bytes, more or less. Note that even though no "Positions" have been recorded in the file, it still uses up memory.

How much "Free" memory remains in the datalogger? _____. About how many position fixes could you store in that amount of memory under the best of conditions? _____ About how many under poor conditions? _____

{__} "Pause" the data recording process. "Resume" it.

{__} Close the file.

{__} *Review the file*: If there are several files in the memory you may scroll through them. The one you just "collected" should be

───────────────

[3] A kilobyte is 1024 bytes, or characters, of information. A character is a letter, digit, or special symbol. This page represents somewhat less than two kilobytes.

the last. Check the name you wrote down on the scrap of paper. Several items of information are available from the file.

```
Review: A010203A
Rover File
Log Rate:   10sec
Positions:    180

#Features        0
#Positions       0
Size:       7.055K
Duration:     1:05
```

{__} Delete the file. (Be careful to delete only the file you created. It will be the one with the highest sequence number; also check that it was made today ("today" in Greenwich, that is).

{__} Run through the checklist in Part 1 in preparation for going into the field.

{__} Turn off the receiver.

## In the Field: Collecting Data

{__} Return to the site of your first data collection effort (PROJECT 1B), where you wrote down position fixes on paper. Once you have considered the factors you previously learned about, regarding good data collection:

{__} Position the antenna where it was before,

{__} Turn on the receiver,

{__} Wait until the receiver is locked onto enough satellites,

{__} Make the appropriate notes on the data collection parameter form, and

{__} *Begin recording data*: On the main menu select "1.Data Capture". Then select "1.Open Rov. File". Be very careful that you don't open a base file. Very unfortunate things happen if you get this one wrong. ("Open Base File" is the option used when the receiver serves as a non-moving station to collect data that may later be used to correct the data obtained by roving stations. Among other concerns, when a base file is collected, the receiver records data from all the satellites in view, not just the four at a time used for normal, 3D fixes, and that uses up memory at a great rate.)

It has been determined that about 180 fixes, if taken over a period of several minutes, are statistically sufficient to achieve whatever accuracy your receiver and the situation will allow.[4] Recall that you set the datalogger to record a point every three seconds, so nine minutes should suffice for the data collection effort. If you have not collected the correct number of points in that length of time it is perhaps because, at times, the receiver lost lock on enough satellites. The reasons for this vary widely, but a principal one is that someone is getting in the way of a signal. The datalogger will record a fix only if DOP requirements, signal-strength requirements, number-of-satellite requirements, and so on, are met.

{__} While data are being recorded, take time to flip to some other screens; recording will continue unimpeded. You can look at File Status and GPS Status. You may change such settings as datum, time display, units, and so on. Recording will continue (provided you keep your head out of the way of the satellite signals).

It does not matter what display settings you use for datum and units in terms of what is recorded in the datalogger memory. Point fixes will be stored in latitude, longitude, and meters above the

---

[4] A longer time period is required if DoD's Selective Availability is turned on. More about that later.

reference ellipsoid. The lat-long numbers will be based on the WGS-84 datum.

{__} Write in your logbook, on a copy of the form provided, the relevant information about this file.

{__} After nine minutes, verify that you have approximately 180 points. If not, figure out why. If you are well-short of the target number of points, consider running the data collection part of the experiment again. Perhaps raising the PDOP to eight (8) will help, although it may degrade the quality of the data you get.

{__} Close the file. Review it. Finish writing in your log. Assuming everything went according to plan, turn off the receiver.

## PROJECT 2B

In this project you will again use the receiver to take fixes and record the coordinates in the datalogger. The difference between this and the previous project (2A) is that you will move the receiver antenna through space in order to record data points along a track. This capability allows you to generate the more interesting features of a GIS: arcs and polygons, rather than simply points.

You may select one of three ways to move the antenna: by foot, bicycle, or automobile. Read the sections below to decide how you want to take data. (In the event that you cannot take moving data, several data files are provided on the diskette that accompanies this book, so you may process those files. Files found there include those taken by automobile, bicycle, airplane, sailplane, cruise ship, and sailboat. I have no data presently from a hot air balloon; if you generate some, please feel free to send it in.)

Generating a series of points while moving provides some challenges.

- Accuracy. We indicated that 180 points at one location provided virtually the best accuracy you could hope to achieve under any given circumstances. In collecting data while moving, of course, you will collect only one point at each location, with the concomitant loss of accuracy.

- Constellation Vacillation. When you set up the receiver at a single point, you can try to optimize the view of the satellites by staying away from obstructions. While moving, you have little opportunity to pick the points at which the receiver calculates a position.

  As the antenna moves along the path to different positions, the receiver, by simply trying to pick the best set of four satellites to calculate positions, may choose different constellations of four satellites, due to signal obstruction caused by canopy, buildings, overpasses, or tunnels. Because of the various errors, which are different from satellite to satellite, a position fix reported by the receiver using a new constellation may be different from where it would have been had the previous constellation been retained. Thus, the position fixes may not follow a clean line, but may jump from side to side of the true path. A second consideration in taking data along a path is that each change in constellation increases the amount of memory used to store data.
- Multipath. Substantial errors may occur if a given radio signal follows two paths to the receiver antenna. This can happen if a part of the signal is bounced off an object, such as a building. The arrival of two or more parts of the signal at different times can confuse the receiver and produce a false reading.

## Taking Data on Foot

While walking, you need to be careful to keep the antenna high enough so that no part of your body impedes the signal. With the version I GeoExplorer, this means holding it high, well out in front of you, so that the internal antenna can have a clear "view" of the sky. With the GeoExplorer II with the external antenna, it is probably also a good idea not to wave the antenna around any more than is absolutely necessary. A pole, with the unit or antenna affixed to it, attached to a backpack is a nice solution.

Set the interval between logged points based on time, not distance, although you might think that taking a point every so many feet or meters would seem like a good idea. The problem is that the receiver will record almost all spurious points that occur, because it is set to record a point that is more than "d" distance away from the last point. If "d" is, say, set to 20 feet, and a multipath error creates a point 25 feet away from the last point, this spurious point will be recorded. (Recall that the receiver actually generates points at more than one per second.) If, instead, you are recording points, say, every 4 seconds, then there is a better chance that a spurious point may be ignored. You may certainly try this both ways, but you will probably get a ragged path with many spikes in it by using distance as the logging interval.

A brisk walking speed is about five feet per second, so you can set the time interval accordingly, depending on the point spacing you want. For this project, if you are walking, select a path, preferably closed, of a mile or two.

## Collecting Data by Bicycle

An economical way to collect data over significant distances along a linear path is to use a bicycle. But there are a number of pitfalls, in addition to the general dangers all bicyclists face, mostly from automobiles. The principal problem relates to the line-of-sight requirement: unless the antenna is positioned far away from your body, or above it, the receiver will consistently lose its lock on one satellite or another.

One solution — not recommended — is to attach the antenna to the highest point on the "you+bicycle" combination, but bicycle helmets are pretty nerdy-looking appliances anyway; when you duct tape a GPS antenna to the top of it the effect is, well, startling — to the point that neighbors come out with cameras to capture the image. One solution is to "shoot the moon" as it were, to give up any pretense of self-respect, and clothe the antenna in one of those "propeller beanies."

Some other solutions:

• have a special bracket made for a surveying pole with the antenna affixed,

- wear a backpack containing the receiver with the antenna on a pole
- ride a tandem bike (a bicycle built for two), and attach the antenna to the second seat). The effect is still pretty ridiculous but let's face it: you aren't going to collect GPS-data on a bicycle while maintaining a significant level of dignity anyway.

## Collecting Data by Automobile

To collect data by automobile you should place the GeoExplorer as far forward on the dashboard, under the windshield, as you can, so that metal of the car blocks signals as little as possible. An external antenna that lets you look at the screen at the same time you are taking data does not normally come with the version I GeoExplorer, but one may be purchased from Trimble's marine division.

One advantage to collecting data by car is that you can use the power supply from the vehicle. However, *remember not to start or stop the engine with the receiver attached to the car's power supply*. The proper order of events is:

1. Start the car engine,
2. Connect the plug to the car's auxiliary power (cigarette lighter) receptacle,
3. Turn on the receiver,
4. When you are ready, start recording data.

When you are through, *undo* the steps above, *in reverse order*.

With automobile data collection, you may use either time or distance as the logging interval. While multipath may still be a problem, the distance you set between logged points is going to be much greater with the car than while walking. Thus, the chance of a point being generated by a multipath event that is greater than the logging interval is reduced considerably.

While riding as a *passenger* in the auto (the driver is supposed to keep eyes on the road) examine the navigation screen occasionally. You will notice that it gives the car's speed within a mile per hour or so, and the direction as well. (You may set the units of display so as to get miles per hour; changing units will not affect data collection, although pulling the receiver back from its place at the front of the dash may.)

## Actual Data Collection

{__} Choose a method of transporting the antenna.

{__} Choose a route — perhaps a closed loop.

{__} Check and set the parameters under "Configuration ~ Rover Options".

{__} Decide on a recording interval.

{__} Start the data recording process.

{__} Take a minimum of 200 points.

{__} Stop recording data.

{__} Complete the Data Collection Parameter Form

{__} Shut off the receiver.

## Back Inside

{__} If you have not already done so, finish filling out the data collection parameter sheet.

{__} Compare the memory used and the number of fixes collected. How many bytes did the average fix require? _____. If this is larger than 25 then it may be that the unit had to repeatedly change constellations, and hence store more data, to maintain the set of four satellites with the best PDOP. While any extended data collection session will involve constellation changes — after all, each satellite is only visible from a given point of the earth for a few hours a day — you will get the best data if you are careful not to force constellation changes by obstructing the signals.

At this point you have recorded spatial data, but it is essentially locked away in the datalogger. Copying it from the datalogger into a personal computer (PC) is the next undertaking. This process is sometimes called **uploading data.**

{__} The first step in transferring data is to make a physical connection between the receiver/datalogger and the PC. Your instructor will have paved the way for this. Probably all you will have to do is connect the datalogger to a cable with a round plug, the other end of which is connected to the computer.[5]

{__} Your instructor will have set up software on a PC in your laboratory. General instructions for the use of the lab and the PC will have been provided. Your machine should show a "DOS[6] prompt" which will be some variation on:

C:\>

{__} In the absence of instructions to the contrary, type the following DOS commands, ending each line by pressing the "Enter" key. The characters "yi" stand for your initials, up to three.

```
CD    \GPS2GIS
MD    DATA_yi
```

---

[5] People have different attitudes about plugging devices into computers. Some feel it is inconsequential whether or not the computer is turned on as long as one is careful to orient the plug properly and not let mismatched pins make contact. Others feel that power to the machine must be turned off to assure that no damage results from making the connection. Please consult with your instructor to see what the policy of your lab is.

[6] Disk Operating System. If you are not familiar with DOS or its tree-structured directories, you should probably take a short crash course or do a bit of reading.

The first command points the computer's operating system to the place that data for this course are located. The second creates a directory named "DATA_yi" that will hold your data. It is only necessary (and possible) to create this directory once.

Trimble has two software packages that operate on data from the GeoExplorer (and Pathfinder Basic). They are GEO-PC and PFINDER. They have a great deal in common, but PFINDER is the more comprehensive. I will base most of my discussion on GEO-PC, and, if you are using PFINDER, you may assume that what is said here applies to it as well. When there are exceptions, they will be described in a box such as this one, or, for short statements, enclosed in square brackets: [ . . . ].

{__} Now, if you are using the GEO-PC software, type the command:

**GEO-PC**

again terminated by pressing the "Enter" key. This causes the program "GEO-PC" to begin executing.

{__} If you are using PFINDER, type the command:

**PFINDER**

again terminated by the Enter key.

You will see the "title page" or "banner" of the software. Prominently displayed will be a graphically simulated "button" with the word "OK" in it. Note that the "O" is underlined. The computer will not proceed until you "press" this button. You may do this in either of two ways.

- Use the mouse to move the arrow on the screen so it is over the button on the screen; press the left mouse key, or
- On the keyboard, hold down the "Alt" key and press the "O" key.

The combination, "Alt-O", is a **"hot-key"** in the software. Throughout the running of the software program you may use hot-keys to make selections instead of using the mouse. Use whichever method seems most convenient in each situation. However, if your computer is so unfortunate as not to have a mouse at all, you should read the software reference manual to see how to simulate a mouse with the Shift, Enter, and arrow keys. Hot-keys do not cover all situations.

{__} If you have not already done so, press "Okay".

The screen of the computer should present you with a menu. Across the top you should see:

Fig. 2-10 — Geo-PC main menu

(One letter in each menu choice is underlined, indicating the hot-key.)

"Plan" presents you with the times you may collect data, given the PDOP and elevation masks you enter. It also shows you the constellations available at various time intervals and the associated

PDOPs. Finally, you may obtain, for points in time that you specify, the azimuth and elevation of each satellite.

"Comm" allows communication between the datalogger and a PC. Primarily, it is the way in which you load files you collected in the GeoExplorer into the memory of the computer.

"DiffCorr" is used to correct point coordinates by using error value estimates taken by another GPS receiver. We cover this topic in Part 4.

"Output" displays data and allows you to generate files for input into your GIS.

"Utils" allows you to select among several utility programs.

"Config" allows you to configure the parameters of the program as you wish.

"Help" provides assistance in the form of text files related to the particular part of the software you are working with.

"Quit" initiates the process of leaving the software program and returning to the DOS prompt.

If you are using PFINDER, the menu is a little different. You don't have "Plan", presumably because you will be using the Windows-based program "Quick Plan" which provides much more information, and screens with graphics in addition. You won't see "DiffCorr" because it is subsumed under "Utils" in PFINDER. You will see two other headings: "Project" and "Filters" — each discussed later. Other main menu items are the same and serve almost the same functions, so please review the section above for descriptions.

The PFINDER main menu looks like this:

Fig. 2-11 — PFINDER main menu

(Again, one letter in each word is underlined, indicating the hot-key.)

When you click on a menu item, or use the hot-key, a second menu drops down, giving you additional choices. You may activate these choices by mouse click or hot-key. Dropdown menus may be hauled back up with a click of the mouse in a blank area of the screen or with the "Esc" key.

{__} Select "Config". Select "Foreground Color". Pick "Blue" and "Okay" the choice.

{__} Select "Background Color". Pick light gray ("Lt. Gray") and "Okay" the choice.

If you are using PFINDER, ignore the text immediately below and read the box that follows.

{__} Select "Default Path". The small rectangular raised areas are buttons, as you know; the indented rectangular areas are **fields** that contain characters.

Two fields are presented on the resulting screen: "Drive" and "Directory". You can move the cursor from one to the other with the Enter key and you may type into the fields. Probably you should use the "C:" drive, and the directory field should contain:

`\GPS2GIS\DATA_yi`

"Okay" your choices. This will direct the software to the place on the computer's hard disk drive where your data are to be stored.

---

If you are using the PFINDER software:

{__} Select "Project"

"Project" is one of those annoying English words that subsume two distinctly different pronunciations and meanings under a single spelling. As a verb, it can mean, for example, shine light through film onto a screen, or "project" a piece of the earth's surface to make a map — that is make a geographic "projection". But "project," as a noun, can mean a focus of activity, time, data, money, and so on so as to accomplish some goal, as in to "launch a project". In the case of PFINDER, it means the latter, although, since the subject in general is geographic data, no one could complain had you guessed otherwise. In any event, the function of the "Project" is to serve as a closet for files.

Once you have selected "Project" a dropdown menu is displayed. Your choices are "New, Edit, Delete, Set Current, Show Current, List, and Help".

{__} Select "New".

Enter the project name, owner, directory, and a comment if you wish. *None of the directory names should be longer than eight characters.* Your Project Directory should be "DATA_yi". Pressing the "Enter" key takes you to the next field. Make sure "Root Directory" indicates "C:\GPS2GIS\", unless specifically told otherwise by your instructor.

{__} Click on "Okay".

{__} Select "Show Current" to be certain you are working with the right project. Click on "Okay" to leave this screen. To remove the dropdown menu, either click a mouse button in a blank area of the screen, or press the "Esc" key.

{__} You have now set up the PC program so it will install your files in the proper place.

{__} Select "Comm" from the PC main menu.

{__} Select the "Communication Port": By your selection here you tell the software which port (plug) on your PC should expect data from the datalogger. Either trace the cable you plugged in to see if it goes to a marked plug, or ask your instructor. "Cancel" is a good way out (of most menus) if you decide you don't want to make any changes, or if you don't know what you are doing. Put a check mark in "Comm Port ___". Press "Ok".

{__} Turn on the GeoExplorer Receiver/Datalogger. Pick "Configuration ~ Communication ~ Port A". The screen should probably look like this, unless your instructor has indicated otherwise.

```
┌─────────────────────────┐
│  - Comm Port A -        │
│  1.Protocol    XMDM     │
│  2.Baud        9600     │
│  3.Parity      None     │
│─────────────────────────│
│  4.Data Bits      8     │
│  5.Stop Bits      1     │
└─────────────────────────┘
```

This sets up the data communication parameters. Probably your PC has been configured to accept data in this form. The local convention may be different, however. If things don't seem to be working, check with your instructor.

{__} From the main menu on the receiver, select "7.Data Transfer". The receiver should tell you that "Comm is Idle".

A NAVSTAR almanac provides a description of approximately where the satellites are at any given moment in time. It is useful for planning a data collection session, and the receiver needs it to know where to begin looking for the satellites to do position finding. So the receiver automatically collects an almanac every time it is turned on, if it can. To collect an almanac, the receiver needs to be receiving signals from at least one satellite for about fifteen minutes. Almanacs are available from the satellites, all day, every day, and usually have validity for up to three months. Of course, if a new satellite is put up, or the existing ones are rearranged in their orbits, the almanac can become useless immediately. In any event, it is probably best to transfer an almanac to the software each day you upload data. Transferring an almanac to the PC is not necessary in order to use the data you have collected, but it is illustrative of the communication process, so we do it next.

{__} Back on the PC, select "Almanac to PC" in the "Comm" menu. A screen should invite you to "Enter Almanac file name". The drive and directory should appear as set up with the default path. The name of the almanac should be today's date, as taken from the PC's clock. The almanac format is

**yymmdd.ssf**

where "yy" is the year, "mm" the month, and "dd" the day. If you know the date the almanac was taken by the receiver, you could enter it (don't forget the "SSF" extension) but the exact date with respect to almanac collection is usually pretty unimportant.

{__} "Okay" the PC screen, then immediately look at the receiver screen. "Comm is Idle" should be replaced momentarily by "Sending Almanac".

{__} On the PC, select "Data Files to PC". You should be presented with a "sub-screen" inviting you to "Check the Files to Load". (If not, any of a number of things could have gone wrong. Recheck the previous steps; if necessary ask your instructor.) The list should include all the files in the datalogger — that is, those you would see by "sequencing through" the "Data Capture ~ Review File" screen on the receiver itself. On the PC, you can use the "Page Up" (or "PgUp") and "Page Down" (or "PgDn") keys to look at the entire list on the computer screen, or you can "mouse" the vertical slider bar and buttons to the right of the list.

{__} Experiment with selecting files to load into the PC. Use the spacebar, or use the mouse to click on each wanted filename. Click there again to deselect it. Try the "All" button. Try "Clear". A check mark by a filename at the time "Okay" is pressed indicates that that file is to be loaded onto the PC's disk in the specified directory. The absence of a check mark indicates that a file is not to be loaded.

{__} Select only the file(s) you collected.

{__} Click OK. The transfer process should begin, with successive bar graphs showing the progress of the data transmission and conversion.

{__} A reminder regarding Base files (which need not concern you) will appear. Click OK.

{__} Turn off the receiver.

{_} Click on "Quit" and confirm that you want to leave the software. You have finished collecting data and have stored it on a PC. In Part 3 you will examine the data. In Part 4 you may experiment with ways to improve the accuracy of the data. And in Part 5 you will install the data in a GIS.

## Data Collection Parameter Form — GPS2GIS

Memory Remaining in the Datalogger _____K

Hours of Charge Left in Battery _____:_____

Location _____

_____

_____

Date _____ Day _____

Filename _____

Time: Start _____

Using (initial): _____ _____ _____ _____

PDOP _____ (HDOP _____ VDOP _____ TDOP _____)

Position Logging Interval: 1 fix per ____ _____

Number of Fixes _____

Time: Stop _____ Duration _____

Using (final): _____ _____ _____ _____

PDOP _____ (HDOP _____ VDOP _____ TDOP _____)

File Size _____K

Memory Remaining in the Datalogger _____K

Hours of Charge Left in Battery _____:_____

Notes: _____

_____

_____

_____

_____

_____

# Part 3

# Examining GPS Data

IN WHICH *we continue our discussion of the theoretical framework of GPS position finding, and practice using PC software to investigate files collected by GPS receivers.*

# Examining GPS Data

## OVERVIEW

### Some Questions Answered

As you read the last two chapters some questions may have occurred to you. And the answers to these questions may generate other questions. Here are some which come up frequently:

**Question #1:** "The captain of the ship of Figure 2-1 had a map showing the locations of the soundhouses. But how does the GPS receiver know where the satellites are?"

A map is a two-dimensional scale model of the surface of the earth. But models can take many forms, including mathematical. Due to the nature of nature, as elucidated by Isaac Newton and Johannes Kepler, the position of a satellite at any time may be predicted with a high degree of accuracy by a few mathematical equations. A satellite orbiting the earth may be modeled by formulas contained in the memory of the microcomputer in the receiver. When the formulas are applied to bodies at the high altitude of the GPS satellites, where they are free from atmospheric drag, the formulas are relatively simple and can predict the position of the satellite quite accurately.

Almost all formulas have a general form, into which specific numbers are "loaded". For example, in an equation of the form

$$Ax + By = z$$

A and B are "parameters" which represent constant numbers that may be inserted in the equation. When A and B are replaced by actual numbers then the equation is only true for certain values of x, y, and z. The receiver carries the general form of the formulas which give the position of each satellite. Before the range readings are taken by the receiver, the satellites will have broadcast the values of their particular parameters so the receiver can complete its equations. Then, by knowing the current time at a given moment (the moment at which the distance reading is taken), the receiver can know where the satellites are.

Actually, the satellite message coming to the receiver antenna is in two parts. **Almanac** information is broadcast to provide close, but not precise, satellite position information. The almanac for all satellites is broadcast from each satellite. Further, each satellite broadcasts **ephemeris** information (which applies to that satellite only), that provides up-to-the-minute corrections. The satellites are not completely predictable in their orbits because of such forces as gravitational pull from the sun and moon, the solar wind, and various other small factors. So the satellites are carefully monitored by ground stations and told their positions; each satellite then rebroadcasts this information to GPS receivers.

**Question #2:** "The captain needed to know exactly what time it was in order to determine his distance from the soundhouse. How is the clock in the receiver kept accurately on GPS time?"

The short answer is that the receiver clock is reset to GPS time by the satellites each time a position is found. Such resetting is necessary because, while the receiver clock is very consistent over short periods of time, it tends to drift over longer periods. (The four clocks in each satellite cost about $50,000 each; the one in the receiver obviously cost a whole lot less, so you can't expect the same sort of accuracy. If you don't use the receiver for a week or two, you may notice a difference of several seconds between the time the receiver displays and true time.)

If you consider "time" as the 4th dimension and accept that it takes one satellite to fix each dimension, then it is clear that four satellites, working in concert, can set the clock and provide a 3D spatial position.

As you may recall from our discussion of the theory of GPS and from the diagrams you examined, only two satellites are really required for a 2D fix and three for a 3D fix. Given that the receiver has only an approximate idea of what time it is, what must be calculated is a 4D fix. So four satellites are required. It is not correct to say that three satellites are used for the 3D fix and the fourth sets the receiver clock. Rather, all of the satellites operate in concert to find the true "position" of a receiver which "drifts" in time and "drifts" in space.

(GPS has had a revolutionizing effect on the business of keeping extremely accurate time. While most of those who use GPS are concerned with finding positions, the system also supplies extremely accurate time signals to receivers whose positions are known with high precision. GPS has made it possible to synchronize clocks around the world.)


**Question #3:** "The soundhouse sent a signal every minute. How often does a satellite send a signal? What is the signal like?"

Actually, each satellite sends a signal continuously, rather like a radio station broadcasts 24 hours per day. The radio station signal could be considered to consist of two parts: a carrier, which is on all the time, and "modulation" of that carrier, which is the voice or music which you hear when you listen to the station. (You probably have detected the presence of the carrier when the people at the station neglect to say or play anything. The carrier produces silence (or a low hiss), whereas if your radio is tuned to a frequency on which no nearby station is broadcasting you will hear static.)

Each satellite actually broadcasts on two frequencies. Only one of these is for civilian use. (The military units receive both.) The civilian carrier frequency is 1575.42 megahertz (1,575.42

million cycles per second). In contrast, an FM radio receives signals of about 100 megahertz. So the GPS radio waves cycle about 15 times as often, and are, therefore, one-fifteenth as long: about 20 centimeters from wavetop to wavetop.

The modulation of the GPS wave is pretty dull, even when compared to "golden oldies" radio stations. The satellites broadcast only "bits" of information: zeros and ones. For most civilian use, this transmission, and the ability to make meaning out of it, is called the "C/A code" — standing for Coarse/Acquisition code. The word "Coarse" is in contrast to the other code used by the satellites: the "P" or "Precise" code. The C/A code is a sequence of 1023 bits which is repeated every one-thousandth of a second.

A copy of the C/A code might look like this:

```
1 0 0 0 1 1 0 1 0 0 1 0 1 1 1 1 0 1 1 0 0 0 1 . . .
```

and on and on for a total of 1023 bits. Then the sequence starts again. The sequence above probably looks random to you — as though you began flipping a coin, recording a "1" each time it came up heads and a "0" for tails. It is, in fact, called a **pseudo-random noise code** — the term "noise" coming from the idea that an aural version of it would greatly resemble static one might hear on a radio. The acronym is **PRN**.

**Question #4:** "How does the receiver use the 0's and 1's to determine the range from the satellite to the receiver?"

The PRN code is anything but random. A given satellite uses a computer program to generate a particular code. The GPS receiver essentially uses a copy of the same computer program to generate the identical code. Further, the satellite and the receiver begin the generation of the code at the same moment in time.

The receiver can therefore determine its range from the satellite by comparing the two PRN sequences (the one it receives and the one it generates). The receiver first determines how much the satellite signal is delayed in time, and then, since it knows the speed of radio waves, how far apart the two antennas are in space.

As an example (using letters rather than bits so we can have a more obvious sequence, and cooking the numbers to avoid explaining some unimportant complications), suppose the satellite and the receiver each began, at 4:00 p.m., to generate one hundred letters per second:

`G J K E T Y U O W V W T D H K . . .`

The receiver would then look at its own copy of this sequence and the one it received from the satellite. Obviously its own copy would start at 4:00, but the copy from the satellite would come along after that, because of the time it took the signal to cover the distance between the antennas. Below is a graphic illustration of what the two signals might look like to the computer in the receiver:

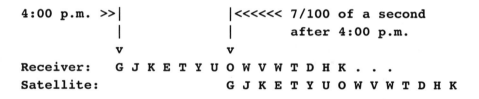

```
4:00 p.m. >>|              |<<<<<< 7/100 of a second
            |              |       after 4:00 p.m.
            v              v
Receiver:   G J K E T Y U O W V W T D H K . . .
Satellite:                 G J K E T Y U O W V W T D H K
```

The receiver would attempt to match the signals. You can see that the signal from the receiver began to arrive seven letters later than 4:00; the receiver's microcomputer could therefore determine that it took 7/100 of a second for the signal from the satellite to reach the receiver antenna. Since the radio wave travels at about 300,000 kilometers per second, the time difference would imply that the satellite was 21,000 (that is, 7/100×300,000) kilometers from the antenna.

**Question #5:** "The receiver must find ranges from at least four satellites to determine its position. How does the receiver "listen to" several satellites at once? Since all satellites broadcast on the same frequency, how does the receiver identify the satellites?"

The first thing to know is that each satellite has its own distinctive PRN code. In fact, the satellite numbers you were logging in the first assignment were the PRN numbers — which is the principle way satellites are identified. A satellite may also have a number painted on its side, but it is the PRN number that counts. When an older satellite is retired, its replacement can take on its PRN number.[1]

Most receivers have at least two or three components, called "channels", which are tuned to receive the civilian GPS frequency. Although all channels are tuned to the same frequency, a single channel can track a GPS satellite by locking onto its PRN code. In more expensive receivers with several or many channels, each channel is assigned full time to tracking a single satellite. Other receivers "time share" a channel — flipping it between satellites, as you might flip between channels on a TV, trying to keep track of two programs at once.

**Question #6:** "What is selective availability? Why do we have to deal with it?"

Selective Availability, or **SA,** is the error deliberately introduced by the GPS managers in the C/A code broadcast to diminish the accuracy of GPS receivers. Sometimes the satellites lie about their positions. Sometimes they lie about when they send the code.

What is the extent of the error caused by SA? The government has guaranteed that 95% of the time a fix will be within 100 meters of the true position. To understand why SA exists you have to realize that the NAVSTAR system started as a military project to provide navigation for units of the armed forces. In the broad sense, GPS was designed as a weapons support system. One doesn't want one's weapons to fall into enemy hands. So steps were taken to deny use of the system to all but authorized receivers. In fact, the system, whose first satellite was launched in 1978, remained secret for several years.

It was probably never planned that you could buy a $300 receiver for your fishing boat. The military fears such uses as a

---

[1] Actually, any satellite could be assigned any PRN number by the managers of the system.

terrorist with a mortar knowing exactly where he was, and hence being able to more accurately target his fire. Or the computer in a missile being able to monitor its position and correct its path during its flight.

If sufficient warning were given, of course, the entire civilian side of NAVSTAR could be shut down to deny its use to hostile forces. (The consequences would be disastrous, but not so much so as a nuclear war.) But, even so, the military is still uncomfortable with allowing the best GPS accuracy in the hands of everybody. As it turns out, very good accuracy may be obtained by using two GPS receivers in concert (we explain this in detail in Part Four) and for the very best accuracy you need two receivers, SA or no. So SA is presently more of a nuisance than anything else. The government announced in March of 1996 that it would be phased out within a decade, perhaps much sooner. The slim protection it provided will be replaced and enhanced by jamming the GPS signals in selected geographical areas, if necessary.

**Question #7:** "How is it that a satellite, cutting Earth's meridians at 55° and moving at 8600 mph, generates a north-south track in the vicinity of the equator?" While the satellite is moving very fast in space, the motion of the corresponding point on the Earth's surface along the satellite's track is about 2100 mph. The satellite's track moves at this speed toward the northeast on the up-swing and southeast on the down-swing, so the eastward part of its motion is in the same direction as the rotation of the Earth. The north or south component of the satellite's velocity is about 1700 mph, while the east component is about 1200 mph. A point on Earth's surface at the equator moves about 1050 miles per hour eastward due to Earth's rotation about its axis. So an observer at the equator would see only a slow drifting of the satellite to the east over the period of an hour or two.

**Question #8:** "If the orbital period is 12 hours, why does each satellite rise and set about 4 minutes earlier each day? Why didn't the NAVSTAR system designers arrange to have the same satellites in view at the same time each day in a given location?"

In answer to the second question, probably the designers didn't have the choice of keeping the satellites arranged so that they showed up in the same place in the sky at the same time — at least not without a lot of extra effort and equipment.

While the satellites are in orbit about the Earth, making their circuits exactly twice a day, the set of satellites are independently in orbit about the Sun. They orbit the sun as a package — sort of like a spherical birdcage that is centered on the Earth that rotates independently within it. The cage does not rotate at all on its axis. (To understand what this means, consider that the moon rotates on its axis once during each trip around the earth, so that it always shows the same face to Earth. If the moon did not rotate, we would see different sides of it as it made its way around the Earth. In contrast the cage *does* present different sides of itself to the sun over the course of a year.) So at any given time (say, noon, when the sun is directly over the meridian you are on), a person on Earth will see (that is, "look through," toward the sun) one portion of the cage on the solstice in January. But from the same point that person would see the opposite side of the cage in July. In effect, then, the cage will be seen from Earth to have made a complete rotation around the Earth once each year. To a person on the Earth, then, the cage apparently moves about 1/365th of a rotation per day. That amounts to about 4 minutes a day — calculated as 1440 minutes in a day divided by the number of days in a year.

**Question #9:** "In earlier text it was suggested that it was somewhat more important that there be a good view of the sky to the south for good reception. Why?" The statement about reception being better towards the south applies only to the middle and upper latitudes in the northern hemisphere. As you know, the satellites are in oblique orbits. Their tracks give the north and south poles a wide berth. In the figure below, you are looking directly down on the north pole at satellite tracks generated over a six hour period. The dashed line is a parallel at 45°, so you can see that there is a dearth of satellites overhead if you go very far north or south.

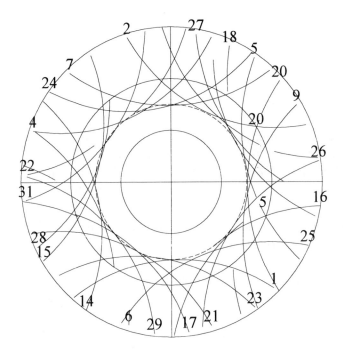

Fig. 3-1 — GPS Satellite tracks looking down on the north pole

# STEP-BY-STEP

*Recall the story of the two "logically challenged" people who rented a boat and went fishing. They were highly successful — catching a lot of fish. Said "A" to "B", "Be sure to mark this spot so we can come back to it." As the day ended and they were approaching the dock, "A" asked "B": "Did you mark that fishing spot?", to which "B" replied: "Sure I did, just like you asked me; right here on the side of the boat." There was a pause as the absurdity of this sunk into "A's" brain. "You idiot! Maybe next time we won't get the same boat!"*

But now, in the last twentieth of the twentieth century, you have a way of marking where you were. In fact, you have done so. In Part 2 you uploaded files from the GPS datalogger into a PC. Our goal now is to look at those files, and some others, both graphically and statistically. You will also acquire an understanding of the quality of your data.

We begin by looking at a file generated by a GPS receiver that circumnavigated Kilauea Caldera, the active volcano on the island of Hawaii.

## PROJECT 3A

{__} Start the GEO-PC or PFINDER software as you did in Part 2. From the main menu, bring up the sub-menu under "Config".

{__} By clicking the left mouse button, set up the following, under "Units".

- Angle: Minutes (the text on the display will give degrees, minutes, and decimal fractions of a minute as the way of designating latitude and longitude)
- Height: Geoid (the text on the display will give height above mean sea level)
- Distance: Meters
- Decimal Accuracy: 3

- Velocity: Meters/Second (though we won't be using it)
- Quadrant: NS/EW (so we don't have to deal with negative longitudes in the western hemisphere).

{__} Okay the choices

{__} Also under "Config", set up the "Coordinate System" as "Latitude/Longitude/Altitude". Once you have done this, the software will ask you to set the "Geodetic Datum". Choose "WGS-84". Note that in these two sub-windows, the item selected when you select "Okay" is the one in the *uppermost* box, by itself. Double check your choices.

{__} Choose "Blue" under "Foreground Color". Choose "Lt Gray" as the "Background Color".

{__} If you are using GEO-PC, choose "Default Path". Make it look like this and okay the screen:

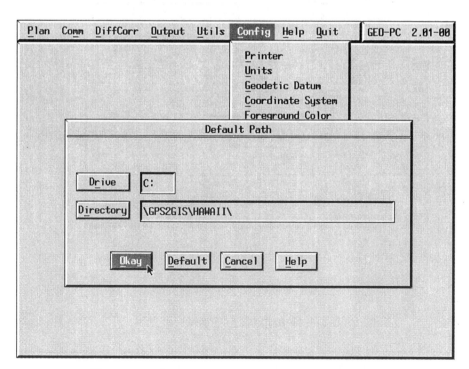

Fig. 3-2 — Setting up a default path in GEO-PC

{__} If you are using the PFINDER software, return to the main menu, choose "Project", and select "Set Current". A list will appear. If it includes "Hawaii" click on that so that it appears by itself in a box at the top of the list. Okay the choice. Now select "Edit" to be certain that "Hawaii" points to the correct directories: \GPS2GIS\ for the "Root Directory" and "HAWAII for the Project Directory. The screen should look like Figure 3-3 below. If it does, okay the screen and then experiment with other choices in this sub-menu.

If the list doesn't include "Hawaii" go back to the sub-menu and choose "New" and make the screen look like Figure 3-2. Okay the screen, then experiment with the other choices in this sub-menu.

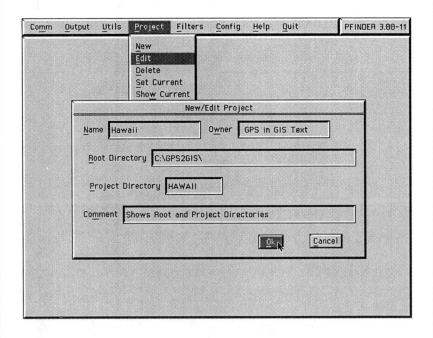

Fig. 3-3 — Setting up a "Project" in PFINDER

{__} *Prepare to display a file graphically on the screen:* Select "Output". Select "Display". You will get a menu that looks something like this:

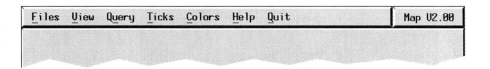

Fig. 3-4 — Menu for display of SSF files

Select "Colors". For background choose "Lt. Gray". (If the background is already set to light gray, you won't be able to see the choice; just cancel the screen in this case.) For highlight choose "Blue".

The previous user may have left the display parameters so that markings called "ticks" will appear on the display. I will show you later how to use ticks, but for now let's just set them "off". From the menu choose "Ticks". Select "Geodetic". A window will appear that contains a box labeled "No Ticks". Put a check in that box, and okay the screen. Now do the same with "Grid" ticks.

{__} *Display the file "VOLCANO.SSF" graphically on the* screen: Select "Files". You should be presented with a window entitled "Enter SSF File Name", showing a drive, directory, and a file specification. The "Drive" and "Directory" are those you specified with "Default Path" (in GEO-PC) [or "Project" in PFINDER]. The "File" will probably be the one the last user worked with. In any event, you want to display the file VOLCANO.SSF so the screen should look like this:

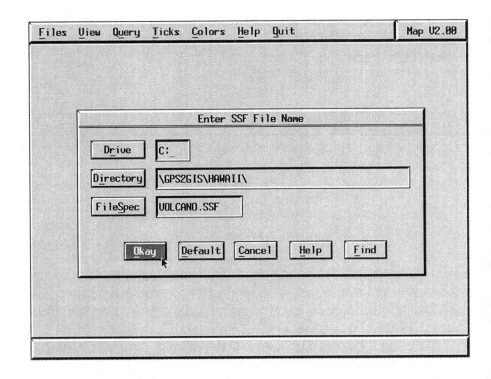

Fig. 3-5 — Providing the SSF filename for display

There are several ways to construct this screen. If you tap the "Enter" key, the cursor (a blinking horizontal or vertical line) moves from field to field. If you simply start typing after moving to a field in this way, the field is cleared except for the characters you type. (This can be somewhat disconcerting if all you wanted to do was change a character or two. If it happens accidentally, just "Cancel" the screen and bring it up again.)

To edit a field, use the "Enter" key (or mouse) to choose the field — and the arrow keys (or mouse) to choose the position in the field. If the cursor is horizontal, the characters you type replace those on the line of text. If the cursor is a vertical line, newly typed characters are inserted. You change the cursor from horizontal to vertical with the "Insert" (or "Ins") key on the keyboard.

If you click on the word "Drive" you may select a different disk drive. If you click on the word "Directory" (don't!) the software will attempt to compile a list of all the directories on the specified drive.

If you click "Default" followed by "FileSpec" or "Okay" you will get a window with a list of all files in the chosen directory. Using the arrow keys, the scroll bar and controls, the keys "PgUp" and "PgDn", and the mouse, you may select the file you want. In this case you want "VOLCANO.SSF". Click on that file_id; note that it now appears in the topmost box.

(In general, if you type a file specification in the "FileSpec" field, and that specification contains DOS wildcard characters ("*" and/or "?"), when you click "Okay" or "FileSpec" you will get a list of all files in the chosen directory that meet that specification. For example, if you know that the file you want has the first three characters as "B12" and that it is an .SSF file, you could type "B12*.SSF" in the field and click "FileSpec". You would be presented with a list of files which would include the one you wanted. You would click on it and okay the choice.)

(Note: There is a "Find" button along the bottom. You may use it if you know some or all of a filename, but don't know which directory it is in. "Find" also finds files meeting the "FileSpec" specification, but *it finds all such files on the hard drive.* There could be thousands of those, if you use the default file specification "*.*". So use this tool carefully, being as specific as possible about the file you want to locate.)

{__} Now that this screen appears as you want it to, okay it. You will get a screen of display options. The appearance of this screen in GEO-PC and PFINDER differ slightly; be sure that the "Display positions" option is checked if it appears, and that "Display features" is *not* checked. "Join" or "Join all points" should be checked as well. This will assure that the points will be shown, and that points that are adjacent in time will have a line between them. Now you have some additional choices to make. Press the "Color" button. Choose "Lt Red" and okay. Press the "Symbol" Button. Choose the "+" symbol and okay. Finally, okay "Display Options" with the button at the bottom of the screen. The track around the volcano will appear, with each position marked.

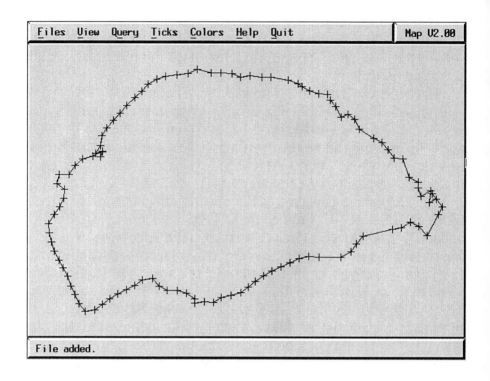

Fig. 3-6 — GPS track around a volcano

{\_\_} "Ticks" are regularly spaced marks placed on the screen for reference. Use them to get an idea of the scale: Choose "Ticks". Choose "Grid". On the screen "Tick Interval - Meters" pick 100. Make the tick color green. Okay.

{\_\_} It seems to be drawing a lot of ticks, no? Press the "Esc" key to stop them. Press the "Yes" button. (If the mouse doesn't work for this use the hot-key combination: Alt-Y.)

{\_\_} Go back into ticks and make the interval a kilometer. Estimate the distance between the most distant parts of the road. _____. Estimate the north-south dimension of the track. _____.

{__} If you are using PFINDER, choose "Measure". Note the bottom of the screen as you move the pointer around the screen. Using the left mouse button, measure the distance from the leftmost point to the rightmost. Measure some other pairs. Experiment with the left mouse button and the right mouse button. What is the maximum distance between any two points? _____ Use "Esc" to leave the measure mode.

{__} Determine the location and time of acquisition of a particular point: Choose "Query". [In PFINDER, then choose "Position"]. Observe the bottom of the screen for instructions and data. Select a position with the mouse, clicking with the left button. Read the position. You should see latitude, longitude, meters above mean sea level, the time the point was taken (in UTC), and whether it is a 2D or 3D point. (In any of the work in this book, if you see a 2D point, it generally means an error has been made in collecting the data.)

{__} The right mouse button will erase the data at the bottom of the screen and allow you to query another point. Do that now.

{__} In which direction was the road traversed? Clockwise( ). Counter-clockwise( ).

{__} To leave the query mode completely, press the right mouse button twice in succession.

{__} Can you find the beginning point of the trip? Determine the approximate location. Then use the "View" dropdown menu and choose "Zoom 2x". Point to the approximate area and press the left mouse button. The area identified will move to the center of the screen and the distance between the points and ticks will be magnified, although the point symbols will not. Zoom in again. Again.

{__} What time did the trip begin? _____ (The answer is in UTC time.) What are the coordinates? _____

{__} Now zoom out using the right mouse button. Again. Again. Use the "Esc" key to leave the zoom mode.

{__} In "View" again, choose "Zoom Normal". Choose "Zoom In". Make a box around an area of interest by using two separate clicks: once in one corner, the second time in the opposite one. Again. Zoom back to full screen view.

{__} Move the image on the screen using "Pan": Place the cross-hairs in the center of the screen, click, move the cross-hairs about halfway toward one corner or one edge of the screen, and click again. Now practice "panning" until you understand the technique. If you lose the image completely, press the right mouse button to end the panning mode, then zoom normal.

{__} Display some statistical information about the VOLCANO.SSF file: Use "Quit" to return to the main menu. Select the utility programs ("Utils"). Choose "Calculate Statistics". You will get a screen, quite similar to the one under "Display", that invites you to enter a filename. It may be that the fields are all filled in properly for you; the software tries to anticipate what you want so it presents you with the file you previously examined. If not, set it up so that you will get VOLCANO.SSF in the \GPS2GIS\HAWAII directory.

{__} Okay the filename selection. Answer "Yes" to the question of forcing calculation of statistics. Note the considerable amount of information on the resulting screen. Not all the text can be viewed at the same time. Use the arrow keys, page keys, and/or slider bars (both vertical and horizontal) to see all the information. If your computer has printing capability, press the "Print" button.

{__} How many points are in the file? _____.

{__} During which GPS week were the data taken? _____.

{__} Approximating from the measurements you took from the display, estimate the average speed, in kilometers per hour, of the car that was carrying the antenna. _____.

{__} Examine the centroid (the 3D average) of the points: Taking "Latitude", for example, the "Mean" is calculated by taking the simple average of all the horizontal values. It approximates the east-west component of the center of the volcano. It is expressed in whatever coordinate system and datum you have chosen. You are also presented with the minimum and maximum values. The standard deviation, which is a measure of how widely dispersed the points are in the east-west direction, is given in the selected units — in this case, meters. The standard deviation doesn't have much meaning for files in which the antenna is moved. However, when a number of points are taken with the antenna in a fixed spot, the standard deviation may be an indication of precision.

{__} Return to the screen on which you selected the VOLCANO.SSF file. (Start with "Output" on the main menu.) Bring up the track of VOLCANO.SSF again if it does not automatically appear. Then click on "Files" again. You get a summary of the selected file(s). To this you want to "Add" another file. From the same directory as VOLCANO, use B021602C.SSF; display it in light magenta, using the simple dot as the display symbol. This track represents the first part of the drive *from* the volcano. What do you have to do to see this entire track? _____.

{__} Add B021603A.SSF in light green, using the same symbol as before. Add B021604A.SSF in yellow. The "Big Island" has two major cities. Towards which one would you say these last two tracks were bound? _____. (You will need an atlas or map.)

{__} The trip *to* the volcano along the same two-lane road is represented by B021520B.SSF. Display this in red. The track should overlay many of the points previously displayed. Use the ticks and the view capabilities to form an idea of the degree to which these tracks are congruent. What (ball park) number would you put on the average of the distance between the tracks? _____.

{___} *Hide the track around the volcano:* Choose "Files". Identify the correct .SSF file and click on it so it appears in the top box. Click "Hide". Now redisplay the file with "Show".

{___} Use "Add" to display VOLCANO.SSF once more, using white and the simple point as a symbol. Zoom up on part of this file as much as you can.

{___} Erase all files from the display: Use "Clear" under "Files".

## PROJECT 3B

{___} Start the software if it is not already running.

{___} Experiment with the "Help" item. Learn how the "slider bars" (one on the right and one at the bottom) allow you to see the text you want. Also experiment with the keyboard "PgUp" and "PgDn" keys. And the keyboard arrow keys, Home key, and End key. Note that you can print the file of instructions should you want a "hard copy." When finished, click on "Okay".

{___} Change the coordinate system to UTM. Change the datum to NAD 27 CONUS. Make certain the units are meters.

{___} *Look at a file of points taken with the antenna stationary over a period of time:* Set the proper DOS path or project, and display (in white, with time-adjacent points joined) the file B022721A.SSF, which you will find in the directory \GPS2GIS\ROOFTOP.

{___} What is the approximate distance between the points farthest apart? _____

For the most part, the time-adjacent points you see here are close together. The track meanders sort of casually around the space. But note that towards the middle of the screen some jagged spikes appear. Zoom up on them. You can see that some time-adjacent points were recorded in widely different positions. Probably what is going on here is that the four-satellite "best PDOP" constellation switches from one set of four satellites to another and then changes back again. This can happen if a satellite is blocked by the operator or some other opaque entity — or if the process of calculating PDOP gives alternating results when the values are very nearly the same. Such spikes make the plot ugly and cause the receiver to use some additional memory for recording the satellites in the new constellation. But in Part 4 we will show you that there is a way to eliminate almost all the variation of the points from the true position value.

{__} How many points are there? _____. Over what period of time was the file collected? _____. What UTM zone contains the points? _____. What city contains the points? _____.

{__} Compare the range (maximum minus minimum) and standard deviations of latitude, longitude, and altitude. Notice that the precision in the vertical direction is considerably worse than in either horizontal direction.

{__} *Make an SSF file that consists of a single point that represents the centroid (the three averages) of all the points in the file:* Under "Utils" select "Averaging Functions". You can use the "Average File" screen to reduce many points in several files to a single coordinate position. We will simply use one file for demonstration purposes: the 900-point file B022721A.SSF.

{__} Click on "Input". Use "Path" to check or set the correct path to the file. "Click "Default", then "Okay" to bring up the list of files in the ROOFTOP directory. Find B022721A.SSF. (I like "PgUp" and "PgDn" to scroll through the list because you get a new box full of filenames with each keystroke. I lose my place otherwise.) Highlight the proper file_id and select the file with the left mouse button or the space bar. (If you wanted to include more files in the calculation you could make checks by them too.) Okay.

{__} The software proposes that you call the output file "FILEMEAN.SSF" but make it instead "ROOFMEAN.SSF". Okay.

{__} Display the original 900-point file; use light cyan. Display ROOFMEAN.SSF in black, using a round symbol. Since the ROOFMEAN.SSF file consists of a single point, what you have here is a graphical representation of the means of the statistics report: the 3D centroid of the 900 cartesian points, though of course you cannot see the altitude dimension. ROOFMEAN.SSF represents the best approximation you can make of the actual location of the antenna, if all you have to go on are the data in B022721A.SSF. In Part 4 you will see how we can make this approximation better, perhaps much better.

## PROJECT 3C

Now that you have a good idea of how to use GEO-PC or PFINDER to display and analyze data taken with a GPS receiver, use this knowledge on the data you collected in Part 2.

{__} Before beginning the computer work:

- Be certain you know the default path [or if in PFINDER, the PROJECT name] you will be working with and the full name(s) of the files you collected.
- Find the detailed map of the area where you took the data so you can compare what appears on the screen with the features shown on the map.

{__} Use "Default Path" [or "Project"] to set up the path to your data in the directory \GPS2GIS\DATA_yi\. (Don't forget the final "\" in the path name.)

{__} Examine the map you used in Project 2A to determine the correct units, coordinate system, and datum to use. Set up the software with these parameters.

{__} Locate the data file you uploaded in Part 2 that corresponds to Project 2A. (Those data are in \GPS2GIS\DATA_yi\.) Use "Output" to display the file. If you type the filename, be sure to type the entire identification, *including the SSF extension.* For example, P052721A.SSF.

{__} When the screen looks the way you want it to, okay it.

{__} The representation now on the screen should be of the points (fixes) you collected, connected in the time order in which they were logged by the receiver.

{__} Select "Query". [In PFINDER also Select "Position".] Click on a fix that is towards the middle of the cluster of points. Read its location at the bottom of the screen and write it down. Is it about where you expected it to be? Check its values against your average value in PROJECT 1C.

{__} Choose "Ticks". Select "Geodetic". Pick a tick spacing appropriate for your data so you get two or three dozen ticks on the screen. (Recall that a minute of latitude is about 1845 meters or 6,053 feet.) Choose a color. Okay. Then turn the ticks off.

## Calculate the Average Position

{__} Return to the main menu. Select "Utils". Select "Calculate Statistics". Check the sub-screen to be sure you are working with the correct file — the one you just displayed. Okay. "Force" statistics calculation. Use the print button to print the file.

{__} Some of the questions below have a second blank. Assuming you took data for both PROJECT 1B and PROJECT 2A in the same location, record the calculations you did in PROJECT 1C. Compare the position fixes.

- What is the average latitude? _____ (What was your Project 1 answer? _____ )
- How many meters in one standard deviation? _____
- What is the average longitude?_____ (What was your Project 1 answer? _____ )
- How many meters in one standard deviation? _____
- What is the average altitude?_____ (What was your Project 1 answer? _____ )
- How many meters in one standard deviation? _____

{__} Compare the position given by the software with the location found on the map. Do they agree? You should expect the average to be within 40 meters or so of the true location horizontally, and 100 meters vertically.

{__} A by-product of running statistics is that a text file, with the information which appeared on the screen, is generated and put into the project directory. Its name is the filename of the points plus the extension "STS". The STS extension stands for "STATISTICS".

{__} Quit the software now and use the DOS print command to print a copy of the "STS" file.

{__} Re-enter the software and set up the datum, coordinate system, and units to correspond to those of the map you used in PROJECT 2B.

{__} Bring in the file of PROJECT 2B and display the path you took while moving the antenna. Now query various points on the display and compare their coordinates with those on the map. Does the display correspond to the actual path you took?

---

{__} If you are using PFINDER, use "Measure" to determine the total length of the segments of your track: _____ .

{__} Choose "Files" and "Add" (with a different color and symbol) the file which contains the points you collected in PROJECT 2A. Unless those points were very near or enclosed by the path of PROJECT 2B, they will not appear on the screen. Choose "View", then "Zoom Normal" to display a screen which contains all points recorded in all files.

## Exercises

Exercise 3-1: Start the software and arrange things so that the default path is \GPS2GIS\EXERCISE\. [In PFINDER, find or make a project that points to \GPS2GIS\EXERCISE.] Display the file EXER3-1U.SSF and determine the general area it represents. Now add EXER3-1S.SSF. (In order to see both these tracks you will have to "zoom normal".) These files are in the vicinities of two major universities on the North American continent. Which ones? Remove the tracks using "Clear" under "Files".

Exercise 3-2: Display EXER3-2H.SSF using dots joined together. What town does the westernmost end of this path represent? Now add EXER3-2N.SSF and EXER3-2S.SSF. At what town do these two tracks join? Now add EXER3-2O. (That's an alphabetic "O".) Zoom up on this last track. What major U.S. city would you expect to find on this circuit?

Exercise 3-3: File EXER3-3S.SSF was generated on a sailboat which went under a newly-constructed bridge. A second file, EXER3-3C.SSF, was generated by a car driving over the bridge. Where did all this take place? What are the latitude and longitude coordinates of the point where the paths of the boat and car cross (use the WGS-84 datum)? What is the name of the bridge? (Hint: use a recent map.) What is the name of the body of water? What major U.S. Air Force base is nearby? When you add file EXER3-3T.SSF, what major city is indicated?

Please see the CD-ROM for additional exercises.

# Part 4

# Differential Correction

IN WHICH *we take a closer look at the subject of GPS accuracy and explore techniques to dramatically reduce errors.*

# Differential Correction

## OVERVIEW

### GPS Accuracy in General

When you record a single position with a good GPS receiver, you are 95% guaranteed by the DoD that the position recorded will be within 100 meters horizontally of the true location of the antenna. Thus, in 5% of the cases, it could be further away.

When a surveyor uses good, survey-grade GPS equipment he or she can locate a point to within a centimeter of its true horizontal position. What are the factors that allow the surveyor to be 10,000 times more accurate than you are? This is a complicated subject. We can cover only the basics in a book of this scope. But you will learn how to reduce errors so that you can record a fix to within two to five meters of its true location. One primary method of gaining such accuracy is called "differential correction".

## Differential Correction in Summary

In a nutshell, the differential correction process consists of setting a GPS receiver (called a **base station**) at a precisely known geographic point. Since the base station knows exactly where it is, it can analyze and record errors in the GPS signals it receives — signals that try to tell it that it is somewhere else. That is, the base station knows the truth, so it can assess the lies being told to it by the GPS signals. These signal errors will be almost equivalent to the signal errors affecting other GPS receivers in the local area, so the accuracy of locations calculated by those other receivers may be improved, sometimes dramatically, by information supplied by the base station.

## Thinking About Error

For the logging of a given point, define **error** as the distance between what your GPS receiver records and the true location of the antenna.

It is useful to dissect the idea of "error". We can speak of error in a *horizontal* plane and differentiate it from the *vertical* error. This is important in GPS, because the geometry of the satellites almost always dictates that the vertical error will be larger than the horizontal.

Another useful distinction is between what we might call **random** error and **systematic** error. It is not easy to make satisfactory definitions of these terms — and, indeed, when you look closely at them they tend to blur into each other. But an example will be illustrative. Suppose we have a machine designed to hurl tennis balls so that they land a certain distance away on a small target painted on the ground. Of course, none of the balls will hit the center of the target exactly; there will always be some error.

What factors might cause errors? The balls are each of slightly different weight; they are not symmetrical and will be loaded into the machine in different orientations; as the temperature changes, the characteristics of the machine may change; the atmospheric pressure and the relative humidity of the air will affect the drag on a ball; and so on. Since it is hard to determine the effects of these factors on the accuracy of the process, we say the factors induce random errors. If there are only random errors in the process, some balls will hit short of the center of the target, some beyond it, some left, some right, and so on. If we shoot 100 balls from the machine we will see a pattern of strikes in the area of target which appears somewhat random, but which clusters around the target.

Now suppose that we had set up our machine and its target when there was no wind, but then a constant breeze of ten miles per hour began blowing from the right across the path of flight of the tennis balls. This would create a systematic error: each ball would land somewhat to the left of where it would have landed in the no-wind condition. We will still see a random pattern of hits, but the average of all hits will be somewhat to the left of the target. This is systematic error; the "system," including the wind, causes it. To correct, we could aim the machine somewhat to the right.

Generally, random errors are those caused by factors we cannot measure or control; systematic errors are those we can account for, measure, and correct for.

## First Line of Defense Against Error: Averaging

When I implied that the surveyor could be 10,000 times more accurate than the average person with a GPS receiver, I was being somewhat disingenuous, mostly for effect. I was comparing a single reading with inexpensive equipment with the average of many readings from expensive equipment. This is not a fair contrast, since you can improve the accuracy of the less expensive equipment by taking many readings at a fixed point. You recall that the strikes of the tennis balls, with no wind, tended to cluster around the target. GPS readings tend to cluster around the true value. We can use the fact that large numbers of random errors tend to be self-cancelling. That is, the average position (take the means of many latitudes, of many longitudes, of many altitudes) will be much closer to the true value than the typical single measurement.

A number of experiments suggest that 50% of the latitude and longitude fixes you obtain with a single receiver operating by itself (i.e., **autonomously**) will lie within 40 meters of the true point. Fifty percent of the altitude fixes will lie within 70 meters. This 50% statistic is sometimes called **Circular Error Probable (CEP).** Ninety-five percent of the fixes will lie within 100 meters horizontally, and 173 meters vertically. These figures include the errors caused deliberately by the managers of the GPS satellite system.

In general, the more fixes you take and the more time you spend, the better your average will be. If you are prepared to take data at one point for several weeks to several months your error will get down to approximately one to two meters.[2] This may not be a practical way to reduce error in most applications.

Another approach which is related to averaging in a different way is to use "over-determined" position finding. As you know, four satellites are required for a 3D fix. But suppose your receiver has access to five or more at a given time. Each set of four of the satellites available will provide a different opinion on the position of the point being sought. A compromise agreement based on all the satellite's input is probably better than the position indicated by any one set of four. The GeoExplorer may be set to collect data in this way.

## Sources of GPS Error

Now look at the specific sources of errors in GPS measurements. Typical error sources and values for receivers of the Pathfinder class are:

| satellite clocks | < 1 meter |
|---|---|
| ephemeris error | < 1 meter |
| receiver error | < 2 meters |
| ionospheric | < 2 meters |

---

[2] According to Chuck Gilbert, of Trimble Navigation, writing in the February 1995 issue of *Earth Observation Magazine.*

    tropospheric            <  2 meters
    selective availability   < 33 meters

These values correspond to averages of many readings rather than the error which might be expected from a single reading. Although experimentation shows that the more fixes you record the better the data become, the increase in accuracy after collecting about 180 fixes at a given location is rarely worth the extra time it takes to record additional fixes. As you will see shortly, there are better ways to get really accurate data.

**Clock Errors.** As you know, the ability of a GPS receiver to determine a fix depends on its ability to determine how long it takes a signal to get from the satellite to the receiver antenna. This requires that the clocks in the satellite be synchronized. Even a small amount of difference in the clocks can make a huge difference in the distance measurements, because the GPS signal travels at about 300,000,000 meters per second.

**Ephemeris Errors.** The receiver expects each satellite to be at a certain place at a particular given time. Every hour or so, in its data message, the satellite tells the receiver where it is predicted to be at time "t" hence. If this ephemeris prediction is incorrect — the satellite isn't where it is expected to be, even by just a meter or two — then the measurement of the range from the receiver antenna to the satellite will be incorrect.

**Receiver Errors.** The receiver cannot exactly measure and compute the distance to each satellite simultaneously. The computer in the receiver works with a fixed number of digits and is therefore subject to calculation errors. The fact is, perfection in position calculation by computer simply is not possible, because computers cannot do arithmetic on fractions exactly. (It is true that other computer operations, such as addition of integers, are perfect.)

**Atmospheric Errors.** For most of its trip from the satellite to the receiver antenna, the GPS signal enjoys a trip through the virtual vacuum of "empty space". Half of the mass of the earth's atmosphere is within 3.5 miles of sea level. Virtually all of it is within 100 miles of the surface. So the signal gets to go the speed limit for electromagnetic radiation for more than 19,000 of the 20,000 kilometers of the trip. When it gets to the earth's atmosphere, however, the speed drops — by an amount that varies somewhat randomly. Of course, since the calculation of the range to the satellite depends on the speed of the signal, a change in signal speed implies an error in distance, which produces an error in position finding.

Significant changes in signal speed occur throughout the atmosphere, but the primary contributions to error come from the ionosphere, which contains charged particles under the influence of the earth's magnetic field, and from the troposphere — that dense part of the atmosphere that we breath, that rains on us, and that demonstrates large variations in pressure and depth.

Really sophisticated GPS equipment can "calculate out" most of the ionospheric error; it considers both the frequencies transmitted by each satellite. Tropospheric errors we seem to be pretty much stuck with, especially using the moderately inexpensive code-based equipment available to civilians.

**Selective Availability (SA).** The lion's share of the "Error Budget" comes from deliberate corruption of the signal by the U.S. DoD. As I have said, they *don't want* the civilian GPS receiver to be able to pinpoint its position. In fact, in the early days, they didn't want the civilian world to know GPS existed. Selective availability is one technique the military can use to keep the system from being too accurate. (Another is called "anti-spoofing".) It appears that the errors are induced by **dithering** (making inaccurate) the signal that tells the ground receiver the exact time. Errors could also be induced by broadcasting a false ephemeris, but it seems that this is not presently being done. Obviously, the military doesn't want to say too much about its techniques for making the good GPS signals "selectively available" — i.e., available only to military receivers.

The DoD has, on occasion, turned selective availability off. (Two notable times were during the Mideast war with Iraq and during the almost-invasion of Haiti.) Under these circumstances an autonomous receiver can supply much more accurate position information. The CEP measure drops to 12 meters horizontally and 21 meters vertically. Ninety-five percent of the fixes fall within 30 meters horizontally, and 52 meters vertically.

## Reducing Errors — Dramatically

As it turns out, most of these errors, including the largest one (SA), can be "calculated out" of the measurements by the process called differential correction. To understand how it works, let me first show the results of an experiment.

The following figure shows two sequences of points (two different files) taken with a GeoExplorer receiver placed at a single location, operating during two different time periods.

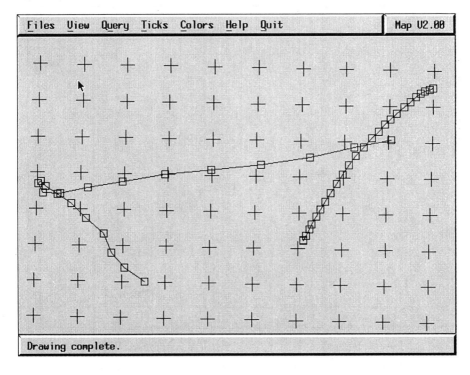

Fig. 4-1 — Two files of the same "point" data (GeoExplorer)

The neat string at right was made of fixes taken about 6 seconds apart; the mess at the left was measured by taking points every 20 seconds during a different time period. (The tick marks are 10 meters apart.) You should realize that each and every fix here is attempting to approximate the same true position. They show up in different positions because of the errors discussed above. We don't know what the true position of the antenna was, but almost certainly each point misses the exact, true position by some amount.

The following figure shows data collected at the same times as in the previous figure; the computations of location were based on the same set of four satellites. The antenna for this receiver (a Pathfinder Basic) was in virtually the same geographic location as the other.

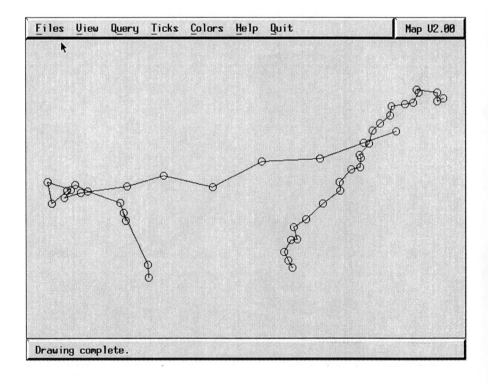

Fig. 4-2 — Same "point" as previous figure (Pathfinder Basic)

Now look at a composite of the two sets of data. The points of the GeoExplorer data are marked with boxes; the Pathfinder data are marked with circles:

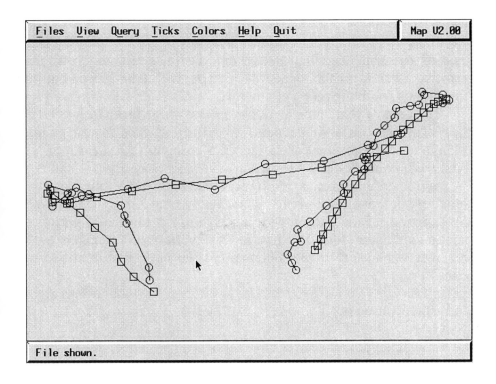

Fig. 4-3 — Composite of four files: same antenna locations

What is interesting is that the two patterns of fixes show remarkable similarity. Pairs of fixes — taken at the same time from the same set of satellites — are in about the same position. This suggests that the **error** for each associated pair of points is almost identical, even though the fixes have come from different receivers and different antennas. And this suggests something which turns out to be true, though I have not proved it here: that receivers calculating from the same signals will suffer from almost the same errors, provided that the antennas are "close". What's close? Two receivers within 500 kilometers (300 miles) will tend to show the same magnitude and direction of errors with respect to the true locations of their antennas, provided the positions are found using the same set of satellites.

## More Formally[3]

The experiment described above demonstrates that much of the error is inherent in the signals — that is, the errors occur before the signals reach the receiver antenna.

To see how that helps us remove most of the error from a GPS fix, let's focus on a single point on Earth's surface (a true point, "T"), and its representation in the GPS receiver (the measured, or observed, point, "O").

Suppose we take a GPS receiver antenna, and place it precisely at that *known point* "T" — a point that has been surveyed by exacting means and whose true position is known to within a centimeter. Now consider that an observation "O" is taken by a receiver connected to the antenna. So we have three entities to consider:

- the position of "T"
- the reading "O", and
- the "difference" between "O" and "T".

---

[3] This discussion provides information about the result of the differential correction process. Actual techniques are involved and varied.

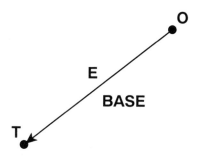

Fig. 4-4 — Error vector – from observed point to true point

We've drawn an arrow from the measured point to the true point. This arrow, which is shown in two dimensions but which would really exist in three, has both a length (called a magnitude) and a direction. An entity that has magnitude and direction is known as a **vector.** We label the vector "E", for "error", because it represents the amount and direction by which the reading missed the true point. Usually we don't know E, but here we can calculate it. The following discussion indicates how.

In general, when we have used GPS, we have used the reported coordinates, "O", as an approximation of "T". The vector "E" was the (unknown) amount by which we missed determining the position "T". As an equation we could write:

$$T = O - E$$

where we record "O" and we disregard "E" to find an approximation of "T". That is, the true coordinates are the observed coordinates minus the error. At best, we could estimate the magnitude — but not the direction — of "E". (It is important to realize that none of these quantities are **scalars** (simple numbers like 23.5) but are three-dimensional entities, so the " – " sign indicates vector subtraction. The concept we are attempting to communicate survives this complexity.)

But if we know "T" exactly, and of course we have the measured value "O", then we can rewrite the above equation to find "E":

$$E = O - T$$

What good is being able to calculate "E"? It allows us to correct the readings of other GPS receivers in the area that are collecting fixes at unknown points.

We have already demonstrated above that if two GPS receiver antennas are close, and use the same satellites, they will perceive almost the same errors. For any given point at any given moment, "E" will be almost exactly the same for both receivers. Thus, for any nearby point reported by a GPS receiver as "o", its true value "t" can be closely approximated simply by applying the equation:[4]

$$t = o - E$$

Since both "o" and "E" are known, the error is effectively subtracted out, resulting in a correct value for "t", as shown by the figure below.

---

[4] Capital "O" and "T" are used to indicate base station variables; lowercase "o" and "t" indicate a rover.

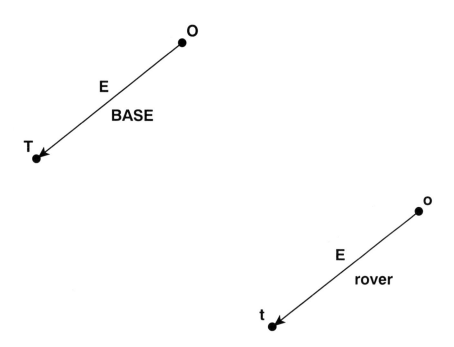

Fig. 4-5 — Known error vector applied to point observed by rover

This technique provides an opportunity for cancelling out most of the error in a GPS position found by an antenna that is close to another antenna which is over a known point. As I mentioned before, "close" is about 500 kilometers or 300 miles. The formula for the amount of error you might expect with differentially corrected data is dependent on the distance between the base station antenna and the rover antenna. A rule of thumb is that the fix will be in error by one additional centimeter for each three kilometers between the two antennas. This relationship is approximately linear: three hundred kilometers would produce error of about a meter.

## Making Differential Correction Work

From a practical point of view, a number of conditions have to be satisfied for the process to work. The base and rover have to be taking data at the same time, and the base has to be taking data frequently. If the base station is to serve the rover, wherever it may be (within the 500 km limit), regardless of when data are taken, the base station must take data from all satellites the rover might see. This can cause a problem: if the base and rover are widely separated the rover might see a satellite that the base cannot view. For example, if the base and rover use the same elevation mask (say 10°), the rover might see "Satellite B" in the figure below, while the base would not.

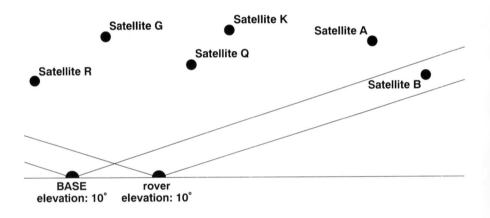

Fig. 4-6 — Base station misses a satellite that the rover sees

Usually base stations are set up with an elevation mask value of 10 degrees. A good rule of thumb is to increase the elevation mask for the rover by one degree for every 100 kilometers (60 miles) it is from the base station. So a setting of 15 degrees works well, for the rover in general if it is within 500 km of the base station. Fifteen degrees is also good for avoiding difficulties due to terrain and that elevation angle reduces errors that tend to occur when signals come from satellites low on the horizon, since those signals must pass through more atmosphere. In any event the rover elevation mask must be set high enough so that there is no possibility the rover will record data from a satellite that the base is not recording, as illustrated by the figure below.

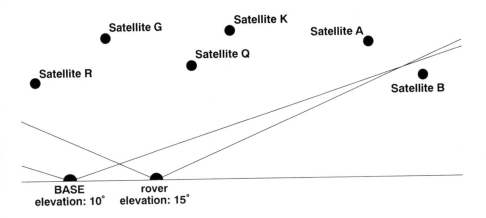

Fig. 4-7 — Neither base nor rover sees Satellite B

Correcting errors by the differential correction method implies that the base station — the receiver at the known point — can communicate with the roving receiver(s). In practice this happens either:

- at the time the readings are being taken, called Real-time Differential GPS (RDGPS), or
- after the readings from the receivers are loaded into a computer, which we call post-mission differential correction.

In RDGPS, a radio link is set up between the base station and the rover. As soon as the calculation for a given point is completed by the receivers, a rover uses the correction signal broadcast by the base station to adjust its opinion of where the point is.

In post-processing GPS or post-mission GPS, the data from both the base and the rover are brought together later in a computer, and the appropriate correction is applied to each fix created by the rover. In the projects which follow, you will use post-mission processing to correct GPS files.

## Proof of the Pudding

The above discussion became pretty theoretical. Is differential correction worth it? You be the judge. The files we used to open this Overview are shown again below. But there is something additional: a "smudge" in the lower left of the figure. Here we added all the fixes that were displayed before, but here now each fix was differentially corrected. In other words, all the fixes with their obvious errors are now within the area of one small circle — presumably the true location of the antennae. The ticks again are ten meters apart, so you can get an idea of the amount of error reduction.

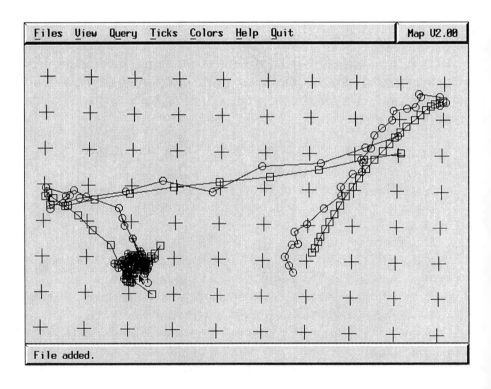

Fig. 4-8 — Eight files: four as collected, then as corrected

# STEP-BY-STEP

## PROJECT 4A

Now that you understand the theory, post-process some sample data. I will describe the process for the GEO-PC software; PFINDER works in much the same way.

{__} In the \GPS2GIS\ROOFTOP directory you will find, using the DOS DIR command, a file named T071122A.SSF. It consists of 334 observations taken within a single-hour period in 1994 with the antenna stationary.

{__} Enter GEO-PC and set the default path (under "Config") to \GPS2GIS\ROOFTOP. [With PFINDER, you may make a project named Rooftop; make the Root Directory "\GPS2GIS\" and the Project Directory "ROOFTOP".]

{__} Under "Config", set the "Background Color" to light gray and the "Foreground Color" to blue. Set the "Coordinate System" to UTM and the "Datum" to WGS-84.

{__} From the main menu, select "Output", then "Display". Choose "Ticks" and set "Grid" so that green ticks will be drawn every 50 meters when a file is displayed.

{__} Now use "Files" to select

\GPS2GIS\ROOFTOP\T071122A.SSF.

Join all points, using a box symbol in blue. It appears that the GPS receiver could not decide where it was, actually. In short, it's a mess! You can see dozens of meters separating widely dispersed points. Approximately how many meters separate the two fixes that are the furthest apart? _____ (Hints and reminders: In PFINDER you may use the "Measure" menu item. In GEO-PC you can estimate from the "Ticks" or use "Query" to get the coordinates and then make a good guess by subtraction; purists may use the Pythagorean theorem.)

{__} "Quit" the display and return to the main menu.

The goal now is to "correct" (i.e., move closer to the true value) each of those points. You will use a file generated by a community base station, only a few miles away, operated by the University of Kentucky Department of Forestry in Lexington. The base station records and keeps the files which contain the errors for all satellites in view (greater than 10 degrees above the horizon). Each hour of tracking activity is recorded as a separate file, beginning on the hour.

In general you can obtain a base station file over a computer communication network such as the "Internet" or the "World Wide Web". Alternatively, you could request a base station file be sent through the mail on a floppy diskette. Base station files are usually large — 150Kb to 250Kb.

The format of each base station file is UYMMDDHH.SSF, where

- "U" indicates "University"
- "Y" is the last digit of the year
- "MM" is the month number
- "DD" is the day number
- "HH" is the hour, in UTC time

{__} What would be the name of the base station file that we would need to correct the file T071122A.SSF? _____
(Hint: What year did I say T071122A.SSF was made?) You will find the base station file also in \GPS2GIS\ROOFTOP.

{__} You may recall that the Trimble software comes with a "hardware key" which prevents a computer from running some of the software options unless the key is present. Differential correction is one of the functions it will not do without the key plugged into the parallel (printer) port of the computer. If it isn't there now, rectify the situation. If you attempt to start differential correction without the key, the software displays the somewhat startling message: "Security device is not present. Requires green dongle Part Number . . ." (You should pardon the expression.)

{__} Now that you have the information from the rover in one file, and the information from the base unit in another, you are ready to correct the rover file.

{__} Now from the main menu choose "DiffCorr" and then "Diff Correction". [In PFINDER choose "Utils", then "Differential Correction".] An intimidating screen with four path and filename blanks appears.

No doubt you understand the requirement for the base and rover files. The "Corrected" file will be the output; it will have the same name as the original file you are correcting, but with a "COR" extension. The _SF file will be explained shortly.[5]

Fill in the missing filenames below.

Base:                    \GPS2GIS\ROOFTOP\U4071122.SSF

Rover:                   \GPS2GIS\ROOFTOP\T071122A.SSF

Difference:              \GPS2GIS\ROOFTOP\U4071122._SF

Corrected:               \GIS2GIS\ROOFTOP\T071122A.COR

{__} Make the screen look like the figure above. Probably the best way is to click on the fields themselves and make the blanks say what you want them to. You may need only to fix the first two (Base and Rover) and tap "Enter". The other two filenames (Difference and Corrected) should be set up automatically. You may also click on the designations on the left of the screen (Base, Rover, etc.); this will allow you to construct the paths and filenames for each file on separate screens.

---

[5] The _SF file extension may show up as "&SF". Don't be concerned.

{__} When you are satisfied that all is correct, okay the screen. After a brief time, you get a report on the process. [In PFINDER, click "okay" to return to the "Measurement Space – Differential Corrections" screen.] Click "Done" (don't click "Okay" — that would start the process over again) to return to the main menu.

{__} Pick "Output", "Display", "Files". If T071122A.SSF does not show up on the screen, display it as you did earlier.

{__} Use "Files" and "Add" to display T071122.COR in light cyan with a circle symbol.
  You should see a noticeable improvement in terms of the clustering of the fixes. Approximately how many meters separate the two fixes of the corrected file that are furthest apart?

_____

  Verify the improvement by looking at the statistics of the two files:

{__} In "Config" choose "Meters", "UTM", and "WGS-84".

{__} In "Utils", "Calculate Statistics" on T071122A.SSF. Write down, or print out, the mean and standard deviation of the northing, easting, and altitude. Now repeat the process for T071122A.COR. Note not only the considerable difference in dispersion, as indicated by the standard deviations, but the considerable difference in the means.
  The three mean values of the COR file constitute the best estimate of the true position of the antenna.
  To review: You just learned to differentially correct an SSF file. What you did is called post-mission processing, or **post processing**, because some time elapsed from the time the file was recorded in the GPS receiver and the time it was corrected by you. Another way to improve GPS data is called "real time differential correction," whereby the base station transmits a radio signal to the rover(s), allowing them to correct their data instantly — in fact, to record the more correct position fixes immediately. We will spend more time discussing this method in Part 7.

## PROJECT 4B

Next we want to post-process a set of several files. The primary difference between these files and the previous one is that these fixes represent a track from a moving antenna, rather than an approximation of a single specific point. Obviously, the overall accuracy of the corrected file will be less, because it would not be appropriate to average the fixes in the file.   You will, however, notice a dramatic improvement in the quality of the data collected by a "roving" receiver.

The files you are going to correct were taken in January of 1994 by both automobile and hiking in the area of Whytecliff Park near Horseshoe Bay, in West Vancouver, British Columbia, Canada. A road map of Vancouver would be useful (are you a AAA or CAA member?) but not necessary.

{__} In the Trimble software, set up the default path to be

\GPS2GIS\WHYTE\

{__} The first file is along the Trans Canadian Highway (the Canadian "Interstate") travelling west from its intersection with 1st Avenue in Vancouver. Display the file T012920B in blue, with a simple dot symbol. Make green grid ticks every kilometer.

{__} Add the file T012921A, in light red. Zoom normal to see it all. Now zoom in on the red road. What happened here, though you can barely tell it from the GPS track, is that the car proceeded from the southeast, went northeast until the road made a large "U", headed southwest and into the parking lot of Whytecliff park. It cruised around the lot, and returned, going northeast, *along the road it had just taken.*  Finally it turned left and encountered a dead end.

{__} Exit the Trimble software. Obtain a DOS prompt in the \GPS2GIS\WHYTE directory. List the files there.  Notice a file named "W4012921.EXE". The name of this file has the format of a community base station file, but the extension is EXE, not SSF.

W4012921.EXE is an interesting type of file. An EXE file is usually a computer program — one that may be executed by the Intel CPU of your computer. This particular file is partly program, but mostly compressed data. It could be called a "self-extracting GPS data file" because when it is executed it generates a community base station SSF file.

{__} Execute "W4012921". You do that just by typing its name at the DOS prompt.

{__} Again list the files in the directory. You will notice that "W4012921.SSF" has been added. This is a standard storage format file from the Whiterock, B.C. community base station, located near the U.S. border in southwestern British Columbia, about 53 kilometers (33 miles) from Horseshoe Bay.

{__} Re-enter the Trimble software. Differentially correct T012921A.SSF with W4012921.SSF. Examine the screen that results from the process. After naming the files, it tells you that 60 of 72 positions were corrected. Below that it says:

edit 0/0/12/0.

This code, consisting of four numbers, tells you about the points that were not moved by the correction process. The first number refers to points in the rover file which were recorded before data collection began on the base station file. The second number is the number of points which were not corrected because of problems with the rover file. The third number, in this case 12, is the number of points that were not moved because of problems with the base station file. (Perhaps the rover used a satellite that the base station did not record.) The fourth number is the number of points in the rover file recorded after the base station file was collected.

{__} While you are using this part of the software, also correct the file "T012921C.SSF". You may note that the correction process goes faster this time — because the computer remembers the corrections from the previous run. That is, "Processing Base" needs to be done only once for each base file, because the &SF (or _SF) Differential File is created, which contains the correcting vectors for the base file. Note this again as you correct "T012921D.SSF". Also note that this base file is not sufficient to correct all the rover files' fixes.

{__} Display T012921A.SSF, again in light red. Set "Ticks" to be drawn every half kilometer.

{__} Display T012921A.COR, in black. Now that is an improvement! Reread the section above that describes the path the car took. It should be easier to follow.

{__} With "Files" and "Hide", eliminate the SSF file, leaving visible only the COR file.
   The receiver was later carried by hikers through the park, from the dead end at the northeast end to the parking lot at the southwest. Two files represent this trip: T012921C goes from the dead end to the pinnacle in the park. T012921D goes from the pinnacle to the parking lot in the southwest.

{__} Display, with a simple point in yellow, T012921C.SSF. The jagged track does not represent where the hikers walked. The spikes in the track probably come from a combination of selective availability and frequent changes in the "best PDOP" constellation, caused by terrain interference. Contributing further to the poor quality of the track was the fact that the receiver was set to record a point every so many meters from the last point, rather than every so many seconds. This has the effect of recording any spurious distant point that is generated. Recall that the receiver calculates a point about every two-thirds of a second.

{__} Continue to trace the hike through the park with T012921D.SSF, also in yellow. Here the GPS receiver was set to record points every 10 seconds. The track looks significantly better.

{__} Display T012921C.COR and T012921D.COR — the first in light blue, the second in light magenta. These tracks aren't perfect

— note some jags and doublebacks — but at least you get a reasonably cohesive picture of the trail. Recall that not all the fixes in the rover file were corrected, which probably accounts for the spurious points remaining.

## PROJECT 4C

Lexington, Kentucky is circumscribed by a (mostly) limited-access ring named New Circle Road. Unlike many cities, Lexington's street pattern resembles a bicycle wheel — spokes and rim — rather than a grid. In November of 1993 a car proceeded from the south, entered the circle road heading west, and proceeded around the city, exiting at every major spoke, making a "U" turn, and re-entering the circle. The primary results of this operation were a GPS file named K110621A.SSF and a police warning regarding illegal "U" turns.

{__} Display the file, which you will find in \GPS2GIS\CIRCLERD, in magenta. The plot should look something like figure 4-9.

{__} To get an idea of the size of the plot set the grid ticks to something appropriate.

{__} Differentially correct the file with U3110621.SSF. Note that the "edit" message indicates that 79 points were not corrected because the base file ended before the rover file did.

{__} Display both the SSF file (in magenta) and the COR file (in light green). Note first that the last eighth of the road has no corrected points. Then zoom up on the south entry point. Use the "Pan" tool in "View" to move westward around the road to the next exit. You can see the track going up the right side of the arterial going north, and again in the right side coming south after the "U" turn. You can also imagine the exit and entrance ramps leading from and returning to the track of the limited access road.

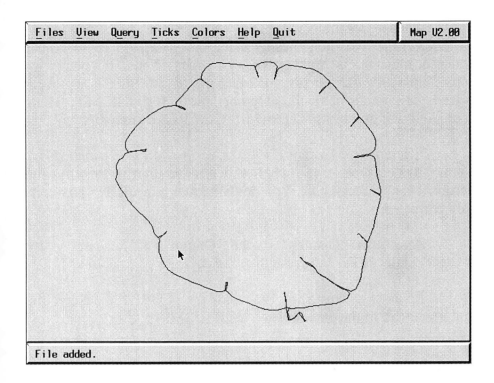

Fig. 4-9 — Trace of fixes around New Circle Road

{__} Continue to pan around the image until you feel proficient with the panning tool.

{__} The way to correct all of a rover file is to get a base file which represents the entire time period over which the rover was taken. Let's look first at the statistics of the rover file to determine its starting and ending times.

{__} Under "Utils" choose "Calculate Statistics". Make certain that the subject file is the SSF one, not COR. Use the "PgDn" key or the mouse and slider controls to see the start and end times. What two base files do you need to cover the entire time period? _____ and _____ .

{__} Choose "Combine SSF Files" under "Utils". Click on "Input" and select the two filenames by using the mouse or the space bar once a filename is highlighted. The files should be placed in chronological order; make sure the appropriate box is checked. Okay the choices. Call the "Output" file U31106HH.SSF. Make certain it will go into the \GPS2GIS\CIRCLERD directory.

{__} Differentially correct K110621A.SSF with U31106HH.SSF. You will be asked if you want to overwrite (that is, destroy) the original K110621A.COR. You do. Note that now all 375 positions are corrected.

{__} Return to "Display" under "Output". If necessary, use "Clear" under "Files" to erase the existing tracks.

{__} Redisplay the SSF file and the COR file. Note the improvement in the extent of the COR file.


**PROJECT 4D**

{__} It may take some research to determine if there is a community base station within 300 miles of your location but GPS equipment vendors can usually supply the information. If you find one, fill out the form at the end of this Part.

{__} Now examine the data you collected in Projects 2A and 2B. What community base station files would you need to differentially correct these files?

 

_____

 

_____

{__} Using the information above, plus instructions from your teacher or employer, obtain the appropriate base station files. Such files are usually kept on hand at the base station locations for two weeks to two months after they have been collected. If the files are EXE files, or "zipped" files, convert them into SSF files.

{__} Use the differential correction process on a file you collected. Display the SSF and COR files with your Trimble software. If the files represent a point, run statistics and compare the means, standard deviations, maxima, and minima of both the SSF and COR files.

## Exercises

Exercise 4-1: In this chapter I have given numbers which indicate the accuracy you might expect from an autonomous GPS receiver. Make a table which puts these numbers together. Your table should indicate horizontal and vertical accuracy, CEP and 95% levels, and whether SA is on or off.

Exercise 4-2: In \GPS2GIS\ROOFTOP you will find two files, C120600G.SSF and C120600G.COR, which represent another attempt to find the location of the coordinates of the rooftop of PROJECT 4A.

{__} Enter the GPS software and set the path to "\GPS2GIS\ROOFTOP\". Set up UTM coordinates: Height as "Geoid", meaning height above mean sea level (MSL), and Datum as WGS-84.

{__} Display these files in different colors with ticks 100 meters apart. Both plots look really terrible, with dispersions of more than half a kilometer. Further, the "corrected" file shows not one, but several clusters of points. What's going on here?

{__} Examine the statistics of each file. Particularly suspicious are the "Altitude" numbers. The standard deviation is essentially zero. And the altitude itself is 47.48 meters below sea level. Since the antenna was at a geographic point almost 300 meters *above* sea level, we may suspect that some altitude setting is the culprit.
 Had I given you the base files necessary to correct these data, you would have discovered the following statements at the end of the process:

Corrected 404 of 411 positions in this pass.
Of these corrected positions, 404 were 2D positions.

So there is the answer. Some dufus (the author) had the GPS unit set on Auto 2D/3D while collecting this particular file. You can only use that setting if you manually and correctly enter the altitude — obviously not something that happened here, since the receiver thought the altitude was –47.5 meters MSL.

The moral of this story is an old one, related to using a computer to turn data into information: Garbage In, Garbage Out.

## Community Base Station Information

Base station name:

_____

Initial character of .SSF files from this base station:

_____

Telephone number:

_____

Account name(s):

_____

_____

Password: (record it elsewhere)

Procedure for downloading files:

_____

_____

_____

_____

_____

# Part 5

# Integrating GPS Data with ARC/INFO

IN WHICH *you learn approaches to converting GPS-software data files to GIS-software data files, practice with sample GPS and digitized coverages, then undertake the process on your own.*

# Integrating GPS Data with ARC/INFO

## OVERVIEW

**To Review:**

You have been given, or have collected, files with the extensions SSF (Standard Storage Format) or COR (for differentially CORrected) files. These files have been processed and displayed using the Trimble GEO-PC or PFINDER software. Your goal now is to use these data files in a GIS where they can be considered with many other data sources that you may have available. This is not a difficult process but you really *must be careful:* it is quite easy to get what looks like a reasonable GIS file, but which has the wrong locations, thus making the activity worse than useless.

I have chosen to illustrate the process using the ARC/INFO software from ESRI (Environmental Systems Research Institute, of Redlands, California) because it is the software that is most widely used by serious GIS professionals, because it is the most comprehensive, and because once data are in ESRI product format they can be converted to many other GIS formats.

ESRI makes several GIS products. Among them are:

- ARC/INFO — an extensive, fully functional GIS which runs primarily on UNIX-based, workstation-class computers.
- PC ARC/INFO — a personal computer (PC) product which is implemented on computers based on the Intel processors (80386, i486, Pentium, and so on). It is a subset of ARC/INFO. There are versions which run in the Microsoft Windows environment and others that run directly under DOS.
- ArcCAD — another PC version of ARC/INFO which runs under AutoCAD (by AutoDesk). Its primary use is providing ARC/INFO functionality for those who wish to operate in the AutoCAD environment.
- ArcView — a windows-based system used primarily for viewing GIS data. An extended discussion of ArcView may be found in Part 6 of this book.

The key to converting from Trimble files to ESRI files is to know that you need only convert from SSF (or COR) files to what is called an ARC/INFO or ESRI **coverage.** That is, once a coverage has been obtained, the entire range of ESRI products is available for your use. You will also have noted that, no matter which Trimble products you use, ultimately you will have files in their standard file format. So it is as though you have a "data tunnel," with wide ranges of products on each side but with the restriction that the data must flow through a narrow passage:

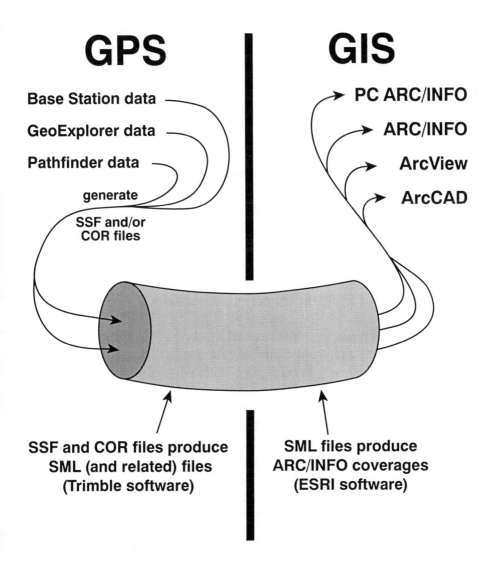

# GPS

**Base Station data**

**GeoExplorer data**

**Pathfinder data**

generate

SSF and/or
COR files

# GIS

**PC ARC/INFO**

**ARC/INFO**

**ArcView**

**ArcCAD**

SSF and COR files produce
SML (and related) files
(Trimble software)

SML files produce
ARC/INFO coverages
(ESRI software)

Fig. 5-1 — Data Tunnel: Trimble GPS Data to ESRI Products

The process of making an ESRI coverage begins with the GEO-PC or PFINDER software. This GPS software will generate a set of files. These files are not ESRI coverage files, but rather are files that ARC/INFO commands will use to create the proper coverage. So you will be executing a multi-step process. The result ultimately will be an ESRI coverage, which is a DOS (or UNIX) directory containing a number of files.

The reason to use a GIS with GPS data is to combine locational data from a variety of sources. So, first and foremost, you must ascertain the parameters of the existing ESRI coverages into which you wish to integrate the GPS data. If you get this wrong, everything will be wrong thenceforth. Among the things you must consider are:

- Geographic datum,
- Projection, if any, used to convert the data from lat-lon representation to a Cartesian coordinate system,
- Units of linear measure, and
- Units of angular measure (i.e., the representation of the graticule on Earth's ellipsoid).

## Prescription for Failure: Incorrect Parameters

- Datum. As you perhaps proved to yourself in Part 1, there may be a difference in the coordinates of a point represented in NAD-27 and WGS-84. In the eastern United States, the values are off in the north-south direction, due primarily to humans learning more about the shape of their Earth. In the Western U.S. the difference is frequently east-west. You probably need only be concerned that you match the datum of the converted GPS file to the datum used by the coverage. You may determine the datum of the coverage in a variety of ways. One that may work is the DESCRIBE command in ARC/INFO. But you should probably carefully investigate the sources of the data and their processing history so you can be certain of the datum used.
- Projection. As you will recall, the 2D components of the Trimble SSF and COR files represent data in the latitude and longitude datum of WGS-84. But large maps shown in such form are (usually) badly distorted visually, so most GIS users elect to store data in some projection or other in which the horizontal and vertical measurements on the map correspond to the horizontal and vertical distances on the Earth's surface. These maps are (generally) much less

distorted. Again, you must tell the Trimble conversion process the correct projection to use.

- Linear Measurement Units. The choices are meters, feet, and survey feet. Survey feet formed the basis of the NAD-27 datum; international feet were used in NAD-83 and WGS-84. What's the difference? Not much, but enough to be significant in some situations. Each comes from a slight distortion of the English unit to make it conform to the metric system. An international foot is based on the idea that there are exactly 0.0254 meters in an inch. A survey foot is based on the equality of exactly 39.37 inches and a meter. If these two conversions were equivalent, you should get exactly unity (1.0000000...) when you multiply them. You won't, and they aren't. What is the product? Use a calculator. _____

- Angular Measurement. Angular measurement units are important only if you are converting a file to a coverage which uses the lat-lon graticule directly. If you do use the graticule, you will want to select degrees, and decimal fractions thereof, because ARC/INFO doesn't utilize minutes and seconds as coordinate values.

The GPS data in the receiver and in SSF files are stored in latitude, longitude, and altitude coordinates. You can, of course, make an ESRI coverage with latitude and longitude directly, as long as you select degrees and fractions of a degree as the output numbers, being sure to use enough decimal places. You may want to do this if you are going to combine the data with other coverages that are stored in that "projection". After all, this is the most fundamental, accurate way. You must realize, however, that any graphic representation of these data will be badly distorted, except near the equator where a degree of longitude covers approximately the same distance as a degree of latitude. Anywhere else, any image of the coverage is visually distorted, and the length of most lines is virtually meaningless, since the "length" is based on differences between latitude and longitude coordinates. Such coordinates do not provide a Cartesian two-dimensional space. (Recall the old riddle: where can you walk south one mile, east one mile, and north one mile, only to find yourself back at the starting point?[1] Not on any Cartesian x-y grid, for sure. Descartes, for his plane, insisted that a unit distance in the "x" direction be equivalent to a unit distance in the "y" direction.)

## The Conversion Process

Once you have determined the parameters of the coverage you want to convert your GPS data to, you are ready to undertake the actual conversion process. In the Step-By-Step section I will lead you through the process, but here I want to point out what the output of the process will be and what you will do with that output to get a coverage.

GPS data are, for our purposes here, simply geographical points or fixes. (You can include attribute data in some Trimble GPS files — a process we describe in Part 7.) An ESRI coverage can consist of:

- a set of individual points. Each point represents a 2D location in some coordinate system.

Fig. 5-2 — Graphics of an ESRI point coverage

---

[1] At the North Pole. And at an infinite number of locations near the south pole, such that you could walk south one mile to a circle around the pole that has a circumference of exactly one mile (or one-half mile, or one-third mile, . . .).

- a set of arcs. An **arc** is composed of a sequence of straight line **segments.** Each segment joins two adjacent (two-dimensional) geographic fixes. At each end of the arc is a geographic fix called a **node.** All the other fixes, which determine the shape of the arc, are called **vertices.**

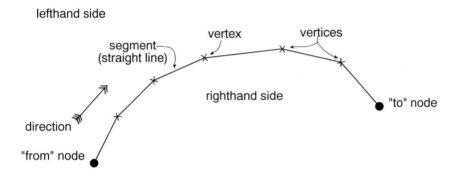

Fig. 5-3 — Graphics of an ESRI arc coverage

- a set of **polygons,** which are closed figures made up of arcs.

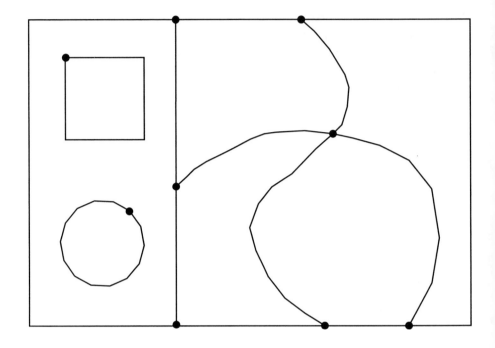

Fig. 5-4 — Graphics of an ESRI polygon coverage

Of course, each coverage may have a feature attribute table (or two) associated with it.

A GPS data file is, of course, simply a set of x-y-z locations, in sequence by the time at which they were recorded. So the GPS data contain the necessary numbers to allow you to make either a coverage which is a set of points, with a Point Attribute Table, or a coverage consisting of a set of arcs with an Arc Attribute Table (AAT). Of course, if you can make arcs you can make polygons.

The Trimble software also allows you to make the files which make polygon coverages directly, but I recommend against it, for a variety of reasons to be discussed later. If you want polygons, make the arcs as separate entities and then use ESRI tools such as ARC/EDIT to put them together into polygons.

## The Files that Generate Coverages

When you run the Trimble software with the "Arc Generate" option, several DOS ASCII files are produced in the directory you specify. To recap, a DOS file identifier (**file_id**) is composed of a filename and an extension. The name may have eight characters, the extension three. This is sometimes called the **DOS 8.3** format. The filename for each of the files generated is the same as the name of the SSF or COR file that was made from the receiver. This name will also be the name of the ESRI coverage which will ultimately be generated. For example, X010203A could be the name of the first file taken on January second during the 3rd UTC hour. Subsequent files produced by the GPS software would be X010203A.SSF and perhaps X010203A.COR. When files are later produced under the "Arc Generate" menu, these generated files have the same name, but differ in their three-character extension. For example, if you specified that each fix in the SSF file was to become a label point in an ESRI coverage, the files you would expect to see produced would be:

X010203A.INF

X010203A.SML

X010203A.GML

X010203A.PTS

Yes, that does seem to be a lot of files, but they give you information and allow you to execute a simple PC ARC/INFO command to get a complete ESRI point coverage, whose name would be X010203A.

The INF file is simply an information file. It might look like this:

```
Input Files:        1
                    1. X010203A.SSF
Information File:   C:\GPS\GEO-PC\DATA\X010203A.inf
Printer Output:     Disable
Data Collected:     Apr 29 17:25:58 1995
GIS:                ARCgenerate
Datum:              WGS-84
Ellipsoid:          WGS-84
Coords:             Universal Transverse Mercator
Projection:         Universal Transverse Mercator
UTM Zone Used:      16S
Units:              METERS 2D
Reference Alt:      2D
Altitude Unit:      N/A
Position Source:    All Positions, LINE
Filter Used:        None

Data Output Files:
------------------------------------------------------------
Stores position information for all points:
C:\GPS\GEO-PC\DATA\X010203A.PTS
Stores position information for all lines:
C:\GPS\GEO-PC\DATA\X010203A.GEN
```

The SML file is an ARC/INFO Simple Macro Language file which, when executed, produces the ESRI coverage. All you have to say (at the ARC prompt) is:

**&RUN X010203A**

and, presto, the coverage appears.

The SML file uses the two other files (the GML file and the PTS file) to produce the coverage. You'll explore this process in more detail when we go through the procedure step-by-step.

# STEP-BY-STEP

## Bringing GPS Data to GIS: Major Steps

A summary of the steps required to produce a GIS coverage from GPS data is as follows:

1. Collect data with GPS receiver.
2. Load data into a PC, making SSF files.
3. Examine data graphically in the PC, and correct it, as appropriate.
4. Convert data from Trimble format to ESRI macro language files using GEO-PC or PFINDER, being especially careful to use the proper parameters.
5. Using ARC/INFO, execute ESRI macro language files.
6. Obtain other GIS data such as TIGER files, commercially available GIS files, and coverages you digitize.
7. Integrate converted GPS data with other GIS data using software modules such as ARCPLOT, ArcView, Overlay, and ARCEDIT.

## PROJECT 5A

In November of 1993 the students and faculty of the Department of Geography at the University of Kentucky participated in a cleanup of the Kentucky River. They took a GPS receiver on their trek; the antenna was mounted on the roof of a garbage scow (née houseboat). One file they collected, along the river from a marina to an island in the river, is C111315A.SSF. Using post-processing differential correction, it was converted to C111315A.COR. In this project you will convert the COR file into an ESRI point coverage using the Trimble software and ARC/INFO.

{__} Start GEO-PC (or PFINDER if you prefer) in the usual way.

{__} You can make life easier if you set up the software so that the default path is \GPS2GIS\RIVER\. [In PFINDER you will have to make a PROJECT to yield this path automatically.]

{__} Select "Output".

{__} Select "GIS". Then select "GIS" from the resulting menu. This second selection will let you identify the GIS software package you are generating data for.

{__} In GEO-PC, to begin to make these data into an ESRI coverage, click on "ARCgenerate". Be certain that this choice appears in a box above the menu window. Okay this choice. [If you are using PFINDER, select "ARC/INFO-DBase".]

{__} Pick "Options" from the GIS main menu, and "Data Type" under that. The safe choices, which you should select, are "4 SVs or more" and "Both Cor & Uncor". Make check marks appear in the appropriate boxes. Now okay your choices. [In PFINDER check the box that says "GPS positions only".]

{__} Select "Elevation to Output" and ensure that elevation data *will not* be exported to the GIS output files. Standard ESRI coverages do not use elevation data.

{__} Select "Coordinate System". Select "Universal Transverse Mercator" and okay the choice.

{__} Because it is so important that you get the datum right when you select a coordinate system, the software now takes you to the "Datum" selection automatically — which is why we skipped it even though it was before "Coordinate System" on the menu. The GIS data that we will be combining our Kentucky River track with was digitized from USGS topographic quadrangles which were based on NAD-27 CONUS (North American Datum of 1927 for the Continental United States). Select this datum and return to the "Options" menu.

{__} Under "Units" click on "Default". Make sure that a check appears next to "Meters". "Decimal Accuracy" can be changed to "2" and "Quadrant" should be "NS/EW". The other settings on this screen are not used in 2D UTM.

{__} The "File" menu item lets you select the SSF and/or COR files that will be used to make the ESRI coverage ("Input Files"). It also lets you determine where the output files will be placed in the directory structure ("Output Path").

{__} Choose "Input Files". Make sure the directory path is

**C:\GPS2GIS\RIVER\**

Put a check by "C111315A.COR" from the menu of files. When you okay the screen a box appears which is titled:

**C111315A.COR --> Arc Info**

The sequence number will say "1", indicating the first (and in this case, only) file in the set of files to be converted into the ESRI coverage. The "User ID" choice is set also to "1", but you may reset it. How this value is reflected in the ARC/INFO coverage depends on other settings in this box. If "All Points" is checked, each fix in the Trimble file is destined to become a point in an ESRI point coverage. The first point will have the value you set in the "User ID" box, and succeeding points will have successively higher integer values.

If "Join as Line" is selected, the "User ID" value will be the ID number assigned to the single arc which the set of fixes in each Trimble file will become. The initial and final fixes will be nodes, while the intermediate fixes will become vertices.

It is not recommended that you use the "Join as Polygon" setting. If it were used, the "User ID" would be the label point of the polygon formed from the file.

{__} Check "All Points". Change the initial "User ID" to "501". Okay.

{__} Pick "Output Path" from the menu. Make certain "Directory" is properly set: \GPS2GIS\RIVER\. When it looks the way you want it, okay it.

{__} Move to the "Info" menu item. Selecting this brings up a summary of the choices that you have made so far. If you can, print this information. In any event, examine it carefully for errors. Okay.

{__} Pick "Run" from the menu. Pick "Preview Graphics". You should see something like this.

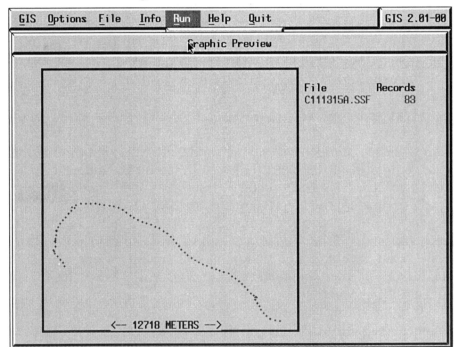

Fig. 5-5 — A COR file about to become SML files

This step is mainly for reassurance. If something isn't right here, you have to backtrack to see where the problem is. A click of the mouse or tapping any key will dismiss this screen.

{__} Select "Execute Conversion". The process of making the ASCII files begins. When the bar graph disappears the process is finished.

{___} Select "View GIS Output". This only allows you to see a text file, but since it lets you look at the coordinates of the points, it is prudent to check it out. The "FileSpec" you should use is "*.PTS". Okay this and you get a menu of files in the directory that have the "PTS" extension. Click on "C111315A.PTS", okay it, and examine the resulting file. You should see point numbers and UTM coordinates such as

```
501, 733618.29, 4195755.69
502, 733546.46, 4195774.28
 +
 +
 +
```

and so on. How many points are there? _____.

{___} Dismiss this screen. Leave the Trimble software by choosing "Quit" twice and answering "Yes". Shortly the DOS prompt will appear.

{___} Change to the directory you specified under "Output Path": In ARC/INFO terminology, this directory will become your workspace. Type:

```
CD \GPS2GIS\RIVER
```

Note that, in DOS, the final delimiter (\) is not included in the pathname.

{___} Get a list of the relevant files in this directory:

```
DIR C111315A.*
```

will reveal the names of the files which the Trimble software just generated. They should be:

```
C111315A.INF
C111315A.SML
C111315A.PTS
C111315A.GML
```

although not necessarily in that order. (You will also see C111315A.COR.)

{__} Examine each of the ARC/INFO-related files by displaying it on the screen: Use a text editor (like EDIT in some versions of DOS) or the DOS "TYPE" command. Use your knowledge of ARC/INFO (or the ARC/INFO manuals) to determine what will happen when you execute the SML (Simple Macro Language) program named C111315A.COR.

{__} Begin the ARC command processor (by typing ARC).

{__} Check to see if there are any ARC/INFO coverages in the directory (with LISTCOVS). If there is a coverage named C111315A, remove it with

```
KILL   C111315A   ALL
```

{__} Execute the macro "C111315A.COR":

```
&RUN C111315A
```

{__} Again list the relevant contents of \GPS2GIS\RIVER:

```
DIR C111315A.*
```

Note that a directory named C111315A has been added to the list of files. This is the ARC/INFO coverage.

{__} List the coverages in the workspace by typing

```
LISTCOVS
or
L   -LC
```

{__} Display an image of the newly created coverage by typing the following commands. After going into ARCPLOT, observe the effect, if any, that each command has on the screen.

```
DISPLAY 4
MAPEXTENT C111315A
TICS C111315A
POINTS C111315A
POINTTEXT C111315A C111315A_ID
QUIT
```

{__} Display the Point Attribute Table of the coverage:

```
LIST C111315A.PAT
```

It is a pretty uninteresting table, but at least you can see the result of placing "501" as the initial point value: it becomes the USER_ID (i.e., C111315A_ID) of the first point collected, with remaining points being numbered sequentially thereafter.

{__} Rename the coverage to KYPTS_yi, where "yi" represents your initials:

```
RENAMCOV C111315A KYPTS_yi
```

(If you do not rename (or remove) the coverage, ARC/INFO will not create the new C111315A coverage in PROJECT 5B. Further, ARC/INFO doesn't tell you it failed to create the coverage. It just pretends it's making a new coverage, but leaves you with the established one.)

{__} After listing coverages to be sure that the renaming command worked, quit ARC/INFO.

## PROJECT 5B

{__} Back in GEO-PC or PFINDER, generate the files for a new coverage named C111315A, but make an *arc* instead of a sequence of points. Basically, you will repeat PROJECT 5A, except this time choose the "Join as Line" option under "File" so the GPS fixes will generate the arc. Instead of the C111315A.PTS file, you will get a C111315A.GEN file.

{__} Use ARC/INFO to make the coverage. After executing the SML file, you will get a coverage with a (very dull: only one record) Arc Attribute Table (AAT) instead of the (hardly exciting) PAT coverage you got before. It does, however, contain the length, in meters, of the GPS arc.

{__} In ARC/INFO, rename the coverage as "KYARC_yi" where "yi" represents your initials:

**RENAMCOV C111315A KYARC_yi**

{__} Change Directory to \GPS2GIS\QUADS.

{__} Execute the LISTCOVS or L -LC command to display the names of the coverages in \GPS2GIS\QUADS. You should find coverages named FORDQUAD and COLEQUAD which contain a few arcs from the FORD, Kentucky and COLETOWN, Kentucky USGS topo quads.[2] These coverages contain some centerlines from major roads on the maps and the banks of the Kentucky River, which flows in the area.

{__} *Start ARCPLOT and graphically overlay the KYARC_yi coverage on the FORDQUAD and COLEQUAD coverages:*

**&REM A text line beginning with &REM indicates a comment**
**&REM for the user. Do not type "comment" lines as**
**&REM instructions to ARCPLOT**

---

[2] It was a short boat trip but representing it involves two quadrangles. Not only are you learning about GPS and GIS, you are also confirming the first law of geography: any area of interest, of almost any size, will require multiple map sheets to represent.

```
DISPLAY 4
MAPEXTENT   COLEQUAD   FORDQUAD   \GPS2GIS\RIVER\KYARC_yi

&REM The following commands show FORDQUAD in green
LINESYMBOL 3
ARCS FORDQUAD

&REM The following commands show COLEQUAD in blue
LINESYMBOL 4
ARCS COLEQUAD

&REM The GPS highway, road, and river tracks are shown
&REM in red
LINESYMBOL 2
ARCS \GPS2GIS\RIVER\KYARC_yi
```

Assuming you properly converted the C111315A.COR file, you should see a red arc, representing the GPS track formed by a GPS receiver that, at a marina, was placed on a boat which journeyed west, then northwest, on the Kentucky River, rounded a bend and stopped at an island.

{__} *Zoom in on the island portion of the display:*

```
MAPEXTENT    *
```

(Use the mouse or cursor keys to make a rectangle around the part of the display which you want magnified. The process of zooming will not be completed until you type the following commands.)

```
CLEAR
```

For your convenience, I've included an ARC/INFO macro called SHOWKYRV.SML. It basically repeats most of the ARCPLOT commands you typed in before to save you the trouble of retyping them.

```
&RUN SHOWKYRV
```

You should see an enlarged version of the river area.

{__} Using the MEASURE command of ARCPLOT, obtain an idea of the dimensions of this part of the Kentucky River:

**MEASURE LENGTH**

(Place the cross-hairs on one bank, click, place the cross-hairs on the other bank, click.)

What are the units of measurement? _____

How wide is the river? _____

What is the length of the trip the boat made? _____

You have successfully combined spatial data from two distinctly different sources: digitized maps and a GPS receiver. Some points:

- Suppose positions had been collected every 30 seconds on the boat trip, the positions were extremely accurate, and the boat stayed precisely in the middle of the river. Would the length of the arc, as found in the AAT, overstate or understate the length of the trip? _____. If there are small random errors in the GPS readings on this trip, what would be the effect of these on the AAT length of the arc? _____. Do these two phenomena tend to reinforce or cancel each other? _____.

- When you executed the SML file using a GEN file as input, you noticed that you got a single arc represented. This will be true, even if the arc crosses over itself. However, if you had such an arc which intersected itself and used BUILD with the LINE option, it would fail. If you CLEAN the coverage, you might get several arcs (and polygons). Without going into detail, let's simply say that you have to be careful when you translate GPS files into arcs. Examine the results carefully and be prepared to edit.

## PROJECT 5C

{__} In the \GPS2GIS\RIVER directory find a GPS file named KYCLAUTO.SSF. This is a track from the marina, north on an access road, north onto I-75 to Lexington's New Circle Road, and west on New Circle.

{__} Convert this file into an ARC/INFO LINE coverage compatible with FORDQUAD and COLEQUAD.

{__} Graphically overlay this coverage with FORDQUAD and COLEQUAD.


## PROJECT 5D

Now comes the real test! Can you integrate GPS data that you collected with GIS data that you digitize? In this project you use the SSF file from PROJECT 2B, in which you moved the antenna through space, and you digitize the general area of the path on a USGS quadrangle.

{__} Obtain the USGS topo quad which contains the area in which you took your GPS data. (If the area spans two or more quads you would have to repeat this procedure for each quad sheet. For the moment, though, just pick the quad sheet which would contain most of your GPS data.) Examine the lower, left-hand corner for the lat-lon coordinates of the southwest corner of the map. Write each coordinate in the following format:

sDDD.MMSS where s is the sign (+ or −), for longitude

sDD.MMSS where s is the sign (+ or −), for latitude

For example, a coordinate (in the U.S.) of west longitude 84 degrees 30 minutes would be represented as "−084.3000".

And north latitude 38 degrees, 37 minutes, and 30 seconds would be "+38.3730".

Your map's longitude in this format: _____

Your map's latitude in this format: _____

You will also need to know the UTM zone number (e.g. 16 ); the zone number should appear on the map. _____.

If you want to offset the northing UTM component you should know the number of meters to *add* to the actual northing value. (Two examples: −4000000, 0.) _____. The GPS data you are working with have no offset, so use zero.

{__} You are about to digitize the portion of a topo quad that indicates some features in the area where you took GPS data. First you will make a blank coverage which contains only ticks in the UTM coordinate system. For simplicity call it TICCOVyi. You may, of course, rename it later.

{__} In DOS, change directory to QUAD2UTM under GPS2GIS:

`CD \GPS2GIS\QUAD2UTM`

{__} There are three ARC/INFO SML macros here as well as an EXE file. Their function is to make a coverage in the UTM coordinate system with ticks properly placed at the corners of the quadsheet. The macro QUADCOV.SML calls the other two SML files and the EXE file, so to begin to make the coverage you need only type:

`&RUN QUADCOV`

You will be informed that the USAGE is

`QUADCOV [cover] [long] [lat]`

To execute QUADCOV you will need the following information:

- The "cover": TICCOVyi

- The longitude and latitude of the *southwest corner* of the 7.5 minute quad, using the formats you wrote out above.

{__} *Run QUADCOV:* Supply your coverage name and the lon-lat designations you wrote out above. You will have to type in the UTM zone number. You will also have to give a code to specify the ticks at the corners of the map.

The goal of this operation is to produce an ESRI coverage consisting of four ticks whose coordinates are the four corners of the map; the values will be proper UTM coordinates in meters. So that map sheets may be tiled together, the tick numbers assigned to each corner are coded to be unique and to reflect the longitude and latitude coordinates in degrees and minutes. Bear in mind that this program and tick numbering system only work for 7.5 minute quadrangles. The breaks between such maps occur at:

```
 0.0 minutes
 7.5 minutes
15.0 minutes
22.5 minutes
30.0 minutes
37.5 minutes
45.0 minutes
52.5 minutes
```

A typical tick number might be 843037522. To determine the position it represents, first look at the number with breaks in the proper places:

```
8430  3752  2.
```

This signifies a tick at 84 degrees 30 minutes longitude, and 37 degrees 52.5 minutes latitude. (The ".5" is assumed; it is not part of the number.) The final digit, "2", indicates the quadrant: The number "1" would mean north of the equator and east of the prime meridian. A digit "2" implies north and west — thus including all of the United States except a bit of Alaska. Number "3" is south and west. Number "4" is south and east.

A quick quiz: What would be the longitude and latitude associated with a tick number of 1220040372? (Hint: group the numbers together, starting from the right. Latitude requires four digits; longitude needs four or five.)

Longitude? _____

Latitude? _____

{__} Use the DESCRIBE command to check the newly created coverage (TICCOVyi). It should have four ticks. Its minima and maxima should include the UTM numbers found on the axes of the map. (If you are not familiar with USGS quadrangle maps and the UTM coordinate system, you may need some help interpreting the map.)

{__} Type: "LIST TICCOVyi.TIC". This will show you the four tick IDs and the UTM coordinates of the ticks.

{__} Using your knowledge of how tick numbers are assigned and the listing generated immediately above, write the ARC/INFO tick numbers for the four corners of the map on the map itself, near each corner.

{__} Mount the map on a digitizing tablet in preparation for digitizing features with ARC/INFO.

{__} Our purpose now is to digitize some features of the quadrangle in the vicinity of the data we took with the GPS receiver. We may want several coverages ultimately, but let's start with one, which we might call "COV1_yi". Start the ADS[3] program with:

```
ADS  COV1yi  TICCOVyi
```

---

[3] If you are more comfortable using ARCEDIT to digitize a map, do so.

This will create a new coverage, COV1yi, which will have the same tick and boundary values as TICCOVyi. While you could digitize directly onto TICCOVyi, you went to a lot of trouble to get its UTM coordinates established, so you probably would like to simply keep it as a "template" in case you need it later.

{__} Put in the tick numbers and the boundary, as usual. The tick numbers are nine- or ten-digit numbers, so extra care is required.

{__} Select features such as roads, streams, powerlines and so on, to digitize. You want to provide a graphical context for the GPS data which you will add shortly. Once you have digitized enough features to orient the GPS data, quit ADS.

{__} Using GEO-PC or PFINDER, convert your GPS SSF or COR file to an ESRI coverage. Call it what you wish: _____. Pay special attention to the "Options" part of the process. You have to get the datum and coordinate system right! (If you are using PC ARC/INFO and the version number is less than 3.5, then the only available transformation to UTM coordinates is to NAD-27. If you are using version 3.5 or later, the programs supplied on the CD-ROM must be changed if you want to use other than NAD-27.)

{__} Using COPYCOV in ARC/INFO, copy the new GPS-based coverage into the \GPS2GIS\QUAD2UTM directory, if it isn't there already. Having all the data in the same directory will simplify the pathnames you have to provide when you display the coverages with ARCPLOT.

{__} Use the approach of PROJECT 5A to graphically overlay the digitized coverage and the GPS coverage. If it appears that everything is in the right place, congratulate yourself.

**Exercise**

{__} To illustrate the importance of getting the correct datum, make a new coverage based on C111315A.COR. What you will do is make this coverage identical to the original point coverage you made, but instead use WGS-84 as the datum. To be on the safe side, copy C111315A.COR to KYWGS_yi.COR and make the ARC/INFO coverage from that file.[4] Once the coverage is created, use ARCPLOT to superimpose this coverage onto FORDQUAD and COLEQUAD as you did before. Note that the boat is no longer in the river. Now add the points from KYPTS_yi, which was based on NAD-27 — the same coordinate system as the map from which the river was digitized. Estimate the distance between the tracks. Note that the difference is much greater in the north-south direction (prominent towards the left of the screen) than in the east-west direction. In this section of the country the UTM coordinates were moved about 200 meters along the meridians and almost not at all on the parallels when the new datum was calculated. Other sections of the U.S. will, of course, show different values.

This is another illustration that while the difference between NAD-27 and WGS-84 may be only a few meters based on latitude and longitude coordinates, it may be hundreds of meters different when based on UTM or state plane coordinates.

---

[4] Make certain that there is no ARC/INFO coverage named KYWGS_yi in the workspace; as you may recall, its presence will prevent a new coverage of this name from being produced.

# Part 6

# ArcView, ArcData, and GPS

IN WHICH *you learn to use a part of ESRI's ArcView software, and use input from the ArcUSA database and GPS files.*

# ArcView, ArcData, and GPS

## OVERVIEW

### Introduction to ArcView

You have seen how to convert your GPS data into ARC/INFO coverages, and, even if you are not very familiar with ARC/INFO, how to make simple maps with those coverages using ARCPLOT. While ARCPLOT provides a very powerful, well-tested, and comprehensive set of mapmaking tools, its format does present problems for the new GIS user. ARCPLOT is a command-driven system whose (admittedly vast) features and capabilities have been developed in a rather additive fashion over the last decade or two. The structure of ARCPLOT's command language, therefore, is less than intuitive, and a good deal of preliminary study is needed to use it effectively.

ArcView, developed by ESRI largely to address these issues, and to maintain its premier spot in the burgeoning desktop mapping market, is a menu-driven system designed for MicroSoft Windows and UNIX platforms. It retains much of the power and flexibility of the various modules of ARC/INFO, but operates in a more user-friendly environment.

ArcView exists in three versions. It is ArcView 2 that is described in this book, but ArcView 3, for the most part, simply enhances ArcView 2. ArcView 1 is somewhat different. It has been put in the public domain by ESRI and is available at no charge. However, to use the original ArcView with the material in this book you would have to make some adjustments. The text has been tested with both the DOS and UNIX versions of ArcView 2.

## Integration of GIS Activities

Besides its user-friendly environment, ArcView serves to integrate many of the activities associated with GIS and desktop mapping. ARC/INFO couples the graphic database with the relational database. ArcView maintains this relationship, and adds to it. Within ArcView is easy and efficient mapping capability, the ability to include charts in the graphic output, and macro language programming support. There are analysis functions as well. ArcView not only handles the standard ESRI coverage, but allows use of the "shapefile," — a digital representation of a geographic area that considerably increases the efficiency with which some operations happen. Planned for ArcView 3 is additional analysis ability, including raster (cell, grid) processing, and network analysis.

Learning to use ArcView is somewhat like learning to ride a bicycle. I can attempt to describe to you how to lean the bicycle into the inside of a corner as you turn the handlebars (ever so slightly, now) and try to sit directly upright with respect to the bike, and . . .

But the truth is that the interactions of the various forces and mechanical components are so complex that you simply have to go through the experiences (even falling off) to truly "understand" the process.

And so it is with ArcView. So I will briefly describe the overall structure of the program and then we will launch into using it, with some GPS data, and some from ArcData.

**The Components of ArcView**

For the GIS professional or student conversant with ARC/INFO, perhaps the greatest hurdle to a quick learning of ArcView is getting used to the nomenclature and the interweaving of the various components. The diagram below may help.

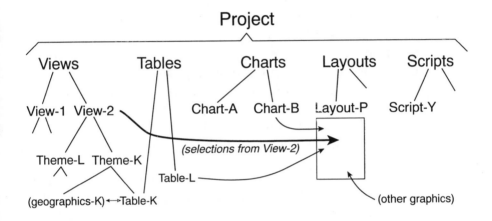

Fig. 6-1 — Portion of a typical ArcView component arrangement

An ArcView **Project,** much like a PFINDER project, constitutes an amalgamation of the data and information associated with a major GIS endeavor. From a graphic point of view, a Project is a window from which a user can access and control the data relevant to the endeavor. From a software aspect, a Project is a single computer file which "points" to other files and sources of information.

Each Project window contains five basic types of components, shown on the second line in the figure above: Views, Tables, Charts, Layouts, and Scripts. The icons for each component type are always available at the side of the "Project" window; each component operates inside its own window, with a set of menus for the particular type. While any number of component windows may be open in a Project, only one of them will be *active* at any given time.

Because this text will place virtually all its emphasis on Views and Tables, I want to describe the Charts, Layouts, and Scripts now, briefly, so we can move on to the issues on which we will spend the most time.

**Scripts** are text files. They contain instructions for the ArcView macro language called **Avenue.** Avenue is used for more advanced applications, or when an organization wants to set up a particular operation for an ArcView operator to do a particular task. You could use ArcView for a long time and not write Avenue scripts. It's the day you notice that you are doing the same operation over and over again, with different data, that you may decide to learn the programming language Avenue, and write a script to save yourself time and avoid mistakes. Avenue is an "object-oriented" programming language, which means that it is capable of handling quite sophisticated data structures, as well as doing the usual complex operations that high-level programming languages do.

**Charts** are just that: graphic representations of data. You may choose regular line graphs, bar graphs, pie charts, and so on. The data for charts comes from tables — essentially those relational database tables that you have used with ARC/INFO. GIS capabilities are being sought by desktop users, who need to be able to integrate such information into presentations. So ArcView provides another tool that lets users present facts and the results of analyses. The products developed by Charts may be integrated into a map, or used by themselves.

**Layouts** constitute the map making capability of ArcView. Here you can combine graphics, "geographics," text, graphs, and other data. Examples include elements you create with the drawing tools provided — ellipses, arbitrary polygons, rectangles. Layouts may contain text in a wide variety of sizes and fonts. Special symbols unique to maps are easily included: north arrows, scale bars, legends, and so on. But mostly, layouts contain geographic information.

As may be inferred from the figure, a **View** is central to work with any data, whether graphic or textual. The fundamental elements of any ArcView session are its views and their corresponding tables. Views are used to display, manipulate and analyze geographic data, which is accessed from various sources such as ARC/INFO coverages, satellite photographs, and tabular data that implies spatial information, such as zip codes and census tracts. The data can read locally from disk or CD-ROM, or across a network from distant data sources.

For those familiar with ARC/INFO, probably the best way to begin with ArcView is with an ESRI coverage. A coverage may become an ArcView theme; **themes** serve as the building blocks of a view. A theme is a set of geographic features which have characteristics in common, very much like a map layer or a coverage in ARC/INFO. So, for example, you might create a land-use view that consists of one theme representing roads, one theme representing soil types, one representing rivers, and one representing sewer lines.

The themes in a view do not actually store geographic data files; rather, they access that data from various sources each time you initiate work on a view containing the theme. The view tools serve only to manipulate and display the source data; therefore any change made to such data is automatically reflected in the themes comprising your view.[1]

In addition to manipulating the display of the themes in your view, you can perform various types of spatial analysis based on relationships within a theme or relationships between themes.

Fundamentally, a theme consists of graphic information about a geographic entity and associated textual (or other) information about that entity. The textual (and other) information is in the form of a relational database table. For example, suppose you have an ESRI coverage that has both a polygon attribute table and an arc attribute table. You may add part or all of this coverage to a View. To add such a theme, you "navigate" to the coverage, and then to the feature type of the coverage. If you select the polygon feature-type, pointers to the graphic representation and the associated PAT are brought into ArcView and become a theme. The name of the theme is listed in the table of contents of the view, and the graphic image may appear in the view window, if you choose. When the theme is active, you may examine and manipulate its table, which

---

[1] This data-reflecting, "dynamic" quality holds for tables, charts and layouts as well. The components of ArcView do not store any data themselves.

you can cause to appear in yet another window.

## Operations on Themes and Views

Once a theme is part of a view, there are many ways to manipulate it. For example, you can:

- select features according to their location by pointing at them with the mouse, or by drawing shapes around them;
- select features on a view by selecting their records from the associated table, according to attribute values;
- select features based on relationships to other features in the theme, or in other themes; and
- change the visual appearance of the features of a theme.

To appreciate the power and ease of use of ArcView, you simply must experience it. You will do that, after just a brief description of one of the data sources we will use in the projects of this part.

## The ArcUSA Database

ArcUSA is a compilation, distributed on CD-ROM, of spatial data related to the U.S. It contains many themes such as roads, lakes, counties, and so on. The data are in the form of ESRI coverages (both DOS and UNIX versions are provided) so they may be used directly and immediately in both ARC/INFO and ArcView. The most detailed data contained on the CD are those that you might expect to find on a 1 to 2 million (2M) scale map. Of course, GIS data are usually referenced by their real world coordinates, so saying that a GIS dataset is of scale "2M" is simply an indication of the level of resolution; GIS data have no scale.

ArcUSA, at the highest level of its data structure, consists of three large directories. The separate 1:25M database serves the need for general information at a small scale, and as an index to the more detailed coverages. The other two directories, at the 1:2M scale, each contain the same data. The most fundamental data, which we will bring into this project, use latitude and longitude, in decimal degrees, as spatial coordinates. The other directory stores data in the Albers Conic Equal-Area projection, which works well for viewing, but not for combining with other data which are not in that projection.

Within each directory are coverages, most of which span the entire country. Some coverages, however, cover only a portion of the U.S., mostly due to their size and level of detail.

# STEP-BY-STEP

## PROJECT 6

### Seeing GPS Data with ArcView

We now turn to using ArcView to examine GPS data together with GIS data from other sources, including the ArcUSA database. In this project, you will bring up a GPS track of a vehicle that travelled from Lexington, Kentucky, south on Interstate 75, west around Knoxville, Tennessee, to the bank of Fort Loudon Lake (the Tennessee River) near Fort Loudon Dam. You will see this track in the context of the counties of Kentucky and Tennessee, major roads and hydrological features. The GPS file you will use first is I090317A.SSF. It was converted to an ARC/INFO coverage of the same name, using latitude and longitude in decimal degrees.

### Starting ArcView

{__} To start the process enter the ArcView program from Windows. Your instructor may have provided information on how to do this.

{__} A window entitled ArcView (across the very top) will appear. Enlarge this window: click on the "up triangle" in the upper right-hand corner of the window. Downsize the window by clicking in the same position (in the box containing two triangles). Now make it full size again.

{__} You want to start ArcView at the most basic level, so you will have to close the "Untitled" window. Either of two ways may be used:

- click on the short, horizontal "bar" in the upper left corner of the window and click on "Close" in the drop-down menu; or
- double click on the "bar".

Be sure not to close the window titled "ArcView" or you will have to wait while it re-initiates.

{__} Close all windows except ArcView. If you are asked if you want to save something, say no. When you are finished, the top of the screen should look like this:

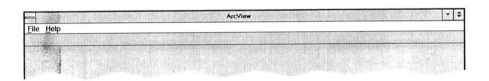

Fig. 6-2 — The most basic ArcView screen

## Getting Help

Getting detailed information about the use of ArcView is facilitated by an "on-line help" feature. Click on the "Help" button. (By the way, "hot-keys" (Alt + a letter) work in ArcView as well as the GPS software.)

The Help system is pretty involved. It is worth spending half an hour or so simply exploring it. Start by selecting "How to Get Help" from the horizontal menu below the title.

Among the features of the help system are:

- A table of contents.
- A searching facility.
- A glossary.

- A "History of Help" topics previously examined in this session, so you can go back to one of them with just a click.
- Several ways of navigating from topic to topic: "Back" (to the previous topic) and << and >> symbols.
- Hypertext. Text strings that are underlined hide other, explanatory text. The screen pointer changes to a hand when placed over this text; when you click on it you get more text. ArcView calls this a "Jump". To highlight all the hypertext-sensitive character strings in a window, hold down Ctrl and press Tab; to highlight each string, in order down the page, repeatedly press Tab.
- The ability to print segments of the Help file.

{__} Exit "Help". Click on the only other ArcView menu item: "File".[2]

## Starting a New Project . . . and Saving It

{__} Click on "New Project". A window with the name "Untitled" will appear.  See Figure 6-3.

{__} *Save this empty project with a new title:* From the "File" menu (note that the menu has changed, now that you have a project opened) select "Save Project As ...". A window of that name appears.

{__} *Put the project into the GPS2GIS directory:* Begin by selecting the "C:" disk drive by clicking in the "Drives:" box. (The little "down arrow" icon associated with the box indicates that options may be chosen.)  Use the slider bar to look at all the available drives. Click on "C:".

---

[2] In ArcView, and Windows in general, there are usually several ways to do any single thing. For example, Alt-F works here instead of "click."

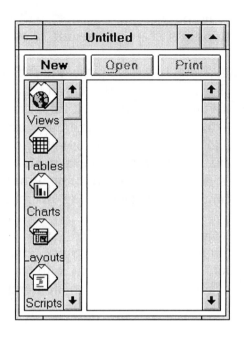

Fig. 6-3 — The ArcView project window

{__} Under "Directories" click on the "C:" folder. This will bring up all the subdirectories under the root. (If they don't appear, double click on the "C:" folder.) You can scroll through them (with the page and arrow keys — the entries are in alphabetical order) to find "gps2gis". (You can make the process faster by typing a "g", which will take you to the top of the entries beginning with "g".) When you locate "gps2gis", double click on it to bring it to the top. It will show up directly under "C:\". The complete pathname, to which you will save the project, will also be shown under "Directories:"

{__} Click in the "File Name:" box. After clearing out any text already there (using delete and backspace), type the name "LXKX_yi" (where yi stands for your initials, up to three) *but don't hit enter.*

Now click "OK". The project "lxkx_yi.apr" will have been saved to the directory C:\gps2gis. ("apr" stands for "ArcView project".)[3]

(It is disconcertingly easy to save a project in the wrong directory. If you just tap "Enter" without the correct name being under "Directories", or if you double click on a folder that is the last one in the list, *ZAP* — the project goes to that directory, correct or not. And since you also double click on folders to see what other folders are in them, it's easy to make a mistake. And it is not particularly easy to delete a project from the wrong directory. But at least you get to make the mistakes using an empty project.)

{__} Close the project window entitled "lxkx_yi". You may be asked if you want to save changes. Since you just saved the project, you do not. You should be back to the basic ArcView window with only "File" and "Help" as menu items.


## Opening a Project

{__} Select "File" and "Open Project", which is the way to bring up a previously saved project.

{__} In the "Open Project" window you go through a procedure much like the one you used to save project LXKX_yi. You select the drive and directory, and filename. In this instance, ArcView probably will have made it easy for you by providing as defaults the drive, directory, and name of the project you just saved. Don't be content with this. Experiment with all the options the window has to offer. Click on the arrow under "List Files of Type" to be certain you will be bringing up files of type "*.apr". Make the drive letter say "a:". (If there is no disk in the "a:" drive, cancel the

---

[3] Note that we use lowercase letters here, for compatibility with the presentation given by ArcView. In DOS, we have stored the file LXKX_yi.APR to the directory C:\GPS2GIS. DOS considers everything as uppercase, regardless of its appearance on the screen. ArcView on a PC may show lowercase letters, but the names used are not case-sensitive.

request.) Then choose "C:" again. Under "Directories" double click on the "C:" folder. Then double click on the "gps2gis" folder. Now in the subwindow under "File Name" you should see (or be able to scroll to) "lxkx_yi.apr". Single click on this, so that it appears directly under "File Name". Now everything is the way you want it, so click "OK".

The project window entitled "lxkx_yi" will appear. Along its left side will be icons called:

Views
Tables
Charts
Layouts
Scripts

{__} The active icon has a box around it and its name is highlighted. Make "Charts" active with a single click. Now make "Views" active again.

{__} Review the ArcView structure figure in the preceding section. Recall that a project may contain views, tables, charts, layouts, and scripts. A view contains themes.

**Initiating a View**

{__} Open a new view: click "New". Lots of things happen. A window appears, titled "View1". Also, two rows of icons appear across the top, just below the menu bar. They are called the "button bar", on the upper level, and the "tool bar", and we will make use of them shortly.

   At this point you have several areas of the screen with which you can control ArcView. To summarize:

- the menu bar (at the very top of the ArcView window)
- the button bar (next bar down)
- the tool bar (down again)

- the icons (arrayed vertically down the left-hand side of the Project window

{__} A window which is active has a colored background in the title bar. You can make a window active by single clicking on the title (or anywhere on the title bar). Make "lxkx_yi" active. Notice that the button and tool bars change. Now make "View1" active.

{__} You can move a window by dragging its title bar. Move "View1" so that it does not overlap "lxkx_yi". A window's size may be changed by dragging a border or a corner of it. Enlarge "View1".

## Views and Coverages: Adding a GPS-Based Theme

{__} *Bring the coverage named i090317a into the "View1"* window: Move the cursor (a pointer) to the menu bar and choose "View". Click on "Add Theme". Look in the C:\GPS2GIS directory for the coverage. Recall that a theme may be an ARC/INFO coverage which might have both point and line features, as well as annotation. It gets a little subtle here. Carefully single click *directly on the file folder* next to the name i090317a. Now single click there again. Note that clicking like this opens and closes the folder. When it is open you can see what is in it. In this case, what is in it is "point" and "annotation". Open the folder and single click on "point". Now click "OK". The "Add Theme" window will disappear and the coverage name "I090317a" will appear in the **Table of Contents** (referred to hereafter as the "T/C") of the View1 window. Also, you will see a fat dot in the T/C, which is the symbol that will be used to plot the points of the coverage (which we will call a "theme" from now on). Click on the little box next to the theme name. A check mark will appear there, and a solid line (composed of points representing the GPS fixes) will show up in the graphic area of View1. A check mark next to the theme name means the theme is to be drawn in the View window.

## Editing the Legend

The dot that represents each GPS fix is so large that it overlaps the adjacent one. We want to use a smaller symbol to represent the individual GPS points. Further, the color of the dot was picked "at random" by ArcView; you will change that also.

{__} From "Theme" on the menu bar, choose "Edit Legend" to bring up a "Legend Editor" window which will let you change the characteristics of a plotting symbol. Double click directly on the circular symbol in the "Legend Editor" window. A window entitled "Marker Palette" appears. (You may want to move this window and enlarge it to see what you are doing.) Click on the paint brush to change the title to "Color Palette". Single click on bright red and click "Apply" in the Legend Editor window. Note the change of color in the display.

{__} In the "Color Palette", click on the "push pin" to change the title to "Marker Palette". Select the square symbol at the top of the middle column. Click in the box across from "Size", type 2.0, then press "Enter". (To type in a box, put the cursor in the field where you want the text. Then use backspace and delete to clear out the existing text. Finally, type the text you want.) Click "Apply" to plot the data with this new symbol size.

If you get confused, click "Revert", which changes things back to the way they were the last time you clicked "Apply". If you get really confused, close (dismiss) the editor and palette windows and start again.

{__} When you have a thin string of red boxes in the view window, dismiss the legend and palette editors.

## Projecting Coordinates

The image on the screen is in lat-lon decimal degrees. You cannot tell that it is distorted, because it is basically a line running from north to south. But when we put other data with it the distortion would become apparent. You will therefore change the view so that it shows data in UTM coordinates.

{__} Make certain the mouse pointer is a simple arrow by choosing the arrow symbol (it is on the tool bar, second from the left). Now slide the pointer around on the view. Note the coordinate values, reported on the right top of the screen, are something like −84.00 (for the east-west coordinate) and 37.00 (for north-south) towards the center of the view. These are the longitude and latitude, respectively, in decimal degrees.

To reduce the distortion, you will convert the picture of the data you see, in the "View1" window, from the spherical coordinate system (the lat-lon graticule based on degrees and decimal fractions of a degree) to the UTM rectangular coordinate system for zone 16. The units will be meters.

{__} From the "View" menu, choose "Properties". Click on "Projection". A "Projection Properties" submenu should appear.

{__} Click in the box labeled "Category" to get a list. Choose UTM. Under "Type" choose Zone 16. (This is the UTM zone which contains the route delineated by the GPS fixes.) Note that the Central Meridian of the zone is 87 degrees west. Okay the "Projection Properties" window. In "View Properties" do not change "Map Units" (from "projected meters") but change "Distance Units" to kilometers. Click "OK".

{__} Again slide the pointer around the view. Note that the coordinates reported now are approximately 750,000 for the horizontal and 4,100,000 for the vertical towards the center of the screen. It is important to realize that the original data have not been converted — they are still in decimal degrees — but their visual representation *in "View 1"* has been changed to the UTM coordinate system.

{__} Click the title bar to make the lxkx_yi.apr window active. Open the "File" menu by a single click. Drag the cursor (with the left button held down) up and down the menu, noting that each highlighted menu item is explained in more detail at the bottom of the screen in a "message bar." When you have finished exploring, choose "Save Project". Again note the bottom of the screen for information.

## Adding a Theme from ArcUSA

{__} Make the "View1" window active. In the "View" menu, choose "Add Theme". Construct the path to a coverage named "Cty2m" by doing the following: Choose the drive "C:". Double click on the C: file folder. Find the "gps2gis" file folder and double click on it. Double click on "arcusa". Double click on "arcusa_d". Double click on "usa_2mg" — the two million scale dataset in geographic (decimal degrees) format.

{__} In the left-hand window, scroll to find the "Cty2m" folder (this represents a subset of the counties in the U.S.). Open the folder by clicking once directly on it. Click on "polygon", then okay the choices. After a brief time, "Cty2m" will appear in the T/C, along with a colored, solid rectangle which indicates that the theme is a polygon (areal) one.

{__} Turn Cty2m on by clicking in the box next to its name. Note that, after a bit, a number of polygons appear, representing counties which surround the GPS track. The GPS track itself does not appear, however. The reason is that it has been covered up by the polygon features.

A nice characteristic of ArcView is that you can specify the order in which themes in the T/C are drawn, and therefore control what themes are painted "on top of" other themes.

{__} *Draw the GPS track on top of the counties:* Reverse the order of the two entries by holding the cursor on the T/C entry "Cty2m", and dragging it "south" of the "I090317a" entry. View1 will be redrawn, showing both themes. The priority for drawing themes is directed by their order in the T/C. First, the theme at the bottom of the T/C list is "painted" on the view. Next to be painted is the theme second from the bottom, and so on. (Generally you would want to paint areal themes first, linear themes next, and point themes last.)

Themes in the T/C must be either on or off (drawn or not). They must also be **active** or not. A theme is active if the area in which its name and symbol appear raised. You make a theme active by clicking on its name. A theme must be active for certain operations on it to take place (e.g., editing its legend).

{__} Make I090317a active. Make Cty2m active — which makes I090317a no longer active. (More than one theme may be made active at a time, but this is a more advanced maneuver. For the moment, know how to make a single theme active, and how to make it visible.)

### Identifying Particular Features of a Particular Theme

{__} With Cty2m active and on, move the pointer to the "**identify tool**" **button,** which is at the extreme left of the tool bar.[4] Click. Note that the button appears to be pushed in. The purpose of the "Identify Tool" is to let you query the table entry for a particular feature *in the active theme.* Move the pointer into the graphic part of the view window, to the county in which the GPS track starts (the northernmost county). Note that the pointer becomes a cross with a little "i" associated with it. Click on the county.

{__} Note that an "Identify Results" window appears. Enlarge it vertically. From the window you can find out what the polygon attribute table (PAT) contains for the county selected. Its area and perimeter are given,[5] its IDs, FIPS (Federal Information

---

[4] Some actions of ArcView are invoked by using items from the various menus. Others are activated by command or tool buttons. And several may be invoked by either method.

[5] The units in the table are those of the data from the coverage (decimal degrees), not the projected units (meters in UTM). Recall that a degree of latitude is about

Processing Standard) codes, state, county, and subregion names. You learn that the county is Fayette, in Kentucky; that the state FIPS is 21, the county FIPS is 67, and the overall FIPS code is 21067, which indicates, if you did not already know it, how an overall FIPS code is assigned.

{__} Click on the county at the end of the GPS track. Loudon county Tennessee? Click on the county just northeast of Loudon. It should be Knox. On the left side of the "Identify Results" window, note that you can examine the information for each of the counties you selected, just by clicking on the entry name. Dismiss the "Identify Results" window.

## Magnifying and Moving the View

{__} Experiment with zooming and panning tools:[6] Make Cty2m the active theme. Under "View" select "Full Extent". You will see all the counties of Kentucky and Tennessee plus one county from each of the other states in the coterminous U.S. (It takes a lot of time to draw all the counties in the U.S., so I've abbreviated the coverage.) Pick "Zoom In" from that same menu. Zoom in again. Each time the view is magnified by a factor of 2. The image at the center of the screen remains at the center.

{__} Select the panning tool (it's the sixth tool button from the left; it looks like a hand); use the hand icon to drag Kentucky (KY) and Tennessee (TN) to the center of the screen.

{__} The fourth tool button from the left (a magnifying glass symbol with a "+" on it) lets you zoom to a rectangle you drag with the pointer. Zoom up on KY and TN.

{__} Make the I090317a theme active. Under "View", "Zoom to (active) Themes". The entire GPS track should be visible, with

---

111 kilometers.

[6] For an explanation of what each tool does, see the text at the bottom of the screen while the pointer is over the tool icon.

some margins around it.

## Selecting Features

Selecting features is an important aspect of ArcView. Features may be selected from a view or a table. The concept is similar to selecting features and records in ARC/INFO with aselect, reselect, and nselect.

{__} Move the cursor over the selection tool (a rectangular icon, third from the left side on the tool bar). Read the description at the bottom of the screen. Selection works only on visible, active themes. Click on this tool. Make the Cty2m theme active. Click in the county that the trip ends in (Loudon). It should turn yellow, indicating that it has been selected. (By the way, you should probably avoid the use of that bright yellow color for themes, since it is the ArcView standard for selected features.)

{__} In the view menu, pick "Zoom to Selected (features)". Loudon county should pretty much fill the view window. (If not, you may not have made Cty2m active.) You should be able to see individual view fixes.

## Adding Water

{__} In the ArcUSA database, find a coverage named "Lak2m". (Hint: choose "Add Theme" from the "View" menu.) Add its polygon features as a theme to the view. Make its color a medium blue. Paint it on top of the counties, but under the GPS track.

{__} Use the magnify tool to zoom up on a rectangle that encloses both the land and water at the end of the GPS track. It appears from the image that the vehicle that carried the GPS antenna must have been amphibious: several points appear off shore.

{__} Use the measuring tool (it looks like a ruler with arrows and a question mark) to determine the distance from the shore to the furthest point on the GPS track. (Click on the tool, then on a point you want the distance from, then double click on the point you want the distance to.) From the message at the bottom of the screen, it appears that the antenna was a third of a kilometer from shore. In reality the vehicle stopped about 100 meters from the water.

Where does the problem lie? The GPS track probably has its usual level of accuracy. ArcUSA data are less accurate, however. They are taken from 1 to 2 million scale maps; a reading of the manual accompanying the data set indicates that lines may be expected to have an error of 1792 meters. That would certainly account for the problem we see with the GPS track and the lake.

{__} Zoom out, using the magnifying glass with the negative sign ("–") in it. With this tool you drag a rectangle with the following characteristics:

(1) The center of the rectangle will be the center of the zoomed-out window.

(2) The area of the drawn rectangle approximates an area that is a smaller portion of the new view, such that the entire existing view would all fit into that drawn rectangle.

{__} Make the GPS track the active theme. Zoom to it. Rearrange and resize the View window so that it occupies the right half of the ArcView window. Narrow the T/C portion of the window by dragging the boundary between the T/C and the graphics to the left.

## Bringing Up a Feature Attribute Table

With the identify tool you were able to see records of an active theme. You may also work with the entire Feature Attribute Table of any given theme.

{__} Make Cty2m the active theme. Under the "Theme" menu, pick "Table". The "Polygon Attribute Table" (PAT) of Cty2m will appear. (Move and resize this table so that it occupies the left half of the ArcView window.)

{__} Use the horizontal and vertical slider bars to examine the contents of the table. Note that there are 293 records (look at the tool bar).

One of the outstanding features of a GIS is that it provides the user with the ability to:

- make queries of a graphical database and get the answers with text; and
- make queries of a textual database and get the answers with graphics.

The following actions will demonstrate how easy this is in ArcView.

{__} Make the "View1" window active by *clicking in its title bar*. (You can make a window active by clicking anywhere in it, but if you click in places other than the title bar, you can produce unintended side effects!)

{__} Use the selection tool and click in the northernmost county that contains the GPS track. Note that it turns yellow.

{__} Make the "Table" window active. Note that one record of the 293 is selected. That record will appear at the top of the window highlighted in yellow. It corresponds to the selected county. What is the county name? _____.

{__} Use the "Find" item in the "Table" menu to bring up a "Find" subwindow. Type "Loudon" (no quotes). Both the record in the table and the graphical representation should become highlighted in yellow.

{__} Find "Knox". What state is it in? _____. Find "Knox" again. What state is it in? _____.

{__} Make the view active. With the selection tool, drag a slim, horizontal box which is contained entirely within several northern Tennessee counties.

{__} Make the table active. Under "Table", pick "Promote" to bring all the selected records to the top of the table.

{__} Under the "Edit" menu item, "Switch Selection". Do it again. Then "Select None".

{__} With the view window active, make I090317a active. Under the "Theme" menu, pick "Table". "Attributes of I090317a" should appear. The table is pretty dull, as are most "Point Attribute Tables". Make that table active. Under the "Edit" menu, "Select All". This should highlight all the records in the table and highlight the symbols showing the GPS track as well. Dismiss this table, but leave the "Attributes of Cty2m" up, and leave the GPS track selected.

## More Complex Selecting

The following actions demonstrate a fairly sophisticated selection technique. You will select the counties that the GPS track traverses.

{__} With the view window active and the Cty2m theme active as well, pick "Select By Theme" under the "Theme" menu. A dialog box appears. Set it up so that it selects features of the active theme that intersect selected features of I090317a. Make this a new set. To see all the selected graphics, zoom to selected features.

{__} Make the GPS track active. Under "Theme", clear selected features, so the track will reappear in red. (Yellow on yellow doesn't provide much contrast.)

{__} *Look at the records of the selected counties:* Make "Attributes of Cty2m" active. (If you click in the body of the window instead in the title bar, you'll lose the proper selections.) How many counties were selected? _____. Verify this graphically. To look at all the records of the selected counties, pick "Promote" under the "Table" menu.

{__} Click on the record of Laurel county Kentucky using the pointer icon. Note how all the other selected records and images disappear. Clear all selections by picking "Select None".

## Other Cool Table Operations

Several other operations may be done on tables. A few examples follow.

{__} Click on the attribute heading "Cnty_name". Note that it is highlighted by changing to white text on black. From the field menu, pick "Sort Ascending". Note the effect on the table.

{__} *Select all the counties of Tennessee:* Begin to "build a query" either by picking the hammer tool or picking "Query" from the "Table" menu. A window appears which contains "Fields" (attribute names), "Values" (attribute elements), and operators such as ">=" and "not". The goal is to type or build the expression:

([State_name] = "Tennessee")

By double clicking on the fields and values, and single clicking on the operators, you can construct this expression. Once it is the way you want it, click on "New Set". The rows in the table which represent Tennessee counties will turn yellow. The graphic representation, "View1", will also show all Tennessee counties as yellow.

{__} Promote the selected rows to the top of the table. Page down through the table. Note that all the Tennessee records are at the top.

{__} *Use "Query" again to add Fayette county Kentucky to the selected set:* Use the delete key and/or the backspace key to remove the previous expression. Start with the parentheses and build the query ([Cnty_name] = "Fayette"). Promote the selected counties.

{__} *Demonstrate the importance of having the correct theme active:* Make Lak2m the active theme. Under the "Theme" menu, pick "Clear Selected Features". Nothing happens, because no features of Lak2m are selected. Now make Cty2m the active theme. Again clear selected features. This time Fayette county is deselected.

{__} Save the project.

By this time you have seen a lot of ArcView — but there is much, much more. If you want to learn more about it, including how it can be used to make thematic maps, do the exercises below.


## Exercises

Exercise 6-1: Make a new set by selecting all Kentucky counties in the table. Notice that 121 records are selected, corresponding to 121 polygons. But Kentucky has only 120 counties, so one county must consist of two polygons. Can you find which one? _____.

{__} Clear the selected set.

Exercise 6-2: ArcView can generate thematic maps — maps that represent the value of some variable by shading or coloring polygons differently. The values used may be taken from a table for the theme. In this exercise you will use the attribute (field) named "Cnty_unemp", for county unemployment.

{__} Expand the view window so it is wider but not so tall. Do the same with the table window. Select the counties of Kentucky from the table. Zoom to selected features. Clear selected features.

{__} Turn off all themes except Cty2m.

{__} Make the view active. With Cty2m active, edit its legend. Double click on the symbol used to color the polygons. The "Theme" will indicate Cty2m. Make the field show "Cnty_unemp". When you do this, the "Legend Editor" will indicate a number of areal symbols, each with a value or range of values beside it. The unemployment for the counties is given by a value ranging from 0 to 6, inclusive.

{__} Click "Classify" and choose "Unique Value". Set the number of classes to seven if you see a place to do so. ArcView may have already determined that you need seven classes. Double click on the top areal symbol. The "Fill Palette" will appear; convert it to the "Color Palette" by clicking the paintbrush; move it so you can work with it and the "Legend Editor" at the same time.

{__} Click "Random", then "Apply". An ugly map will appear, with counties shown in wildly different colors. While you could determine the unemployment value in each county by looking at the legend, you don't get an intuitive idea of the progression of the values.

Try a different approach. You want to assign the "color" clear to the zero values (which appear in the field for states other than Kentucky). You want a light pink for counties with a value of "1" for "Cnty_unemp". And you want the color to get increasingly darker for counties with higher unemployment.

{__} Click first on the topmost areal symbol next to the zero. Then assign it the "clear" value by clicking on the clear square (with the "X") in the Color Palette. Next pick the areal symbol next to the "1". Assign it a light pink from the color palette. Assign the symbol that is next to the "6" a purple.

{__} Choose "Ramp", followed by "Apply". A map should appear which shows a progression of colors based on the value of the "Cnty_unemp" field. The counties of other states should appear as clear.

{__} Dismiss the Cty2m attributes table. Zoom to the extent of the GPS track.

{__} Add the ArcUSA coverage RDS2M as a theme. (You will find it in the GPS2GIS directory.) Edit the legend so it shows up in medium purple with a linear symbol of significant width — one is available which even looks a bit like a road with a dividing line.

{__} Bring up the table for Rds2m. Select Interstate I-75, which usually shows up in the table as a "75" in the fields "Inter_rte1". But sometimes the same stretch of highway goes by two different names. When I-75 is in common with another interstate highway designation, the "75" will appear in the table under the heading "Inter_rte2". To get all of I-75 you must include both column headings in your selection process. (Hint1: Use the "or" operator.) (Hint2: Or use "Add To Set".)

{__} Zoom up and pan along the route to see how closely the GPS track corresponds to the selected segments of I-75, particularly in the area around the Kentucky-Tennessee border. Measure the distance from the GPS track to the interstate.

{__} Add the linear (arc) theme \GPS2GIS\R030917A to the view. This is a GPS track based on a different receiver on the same trip as I030917A. The GPS points were converted to an ARC/INFO line coverage. Make it a medium gray line of size 2.0. Ensure that I030917A is painted on top of it.

Notice that the two GPS tracks seem to be almost congruent, whereas I-75 only roughly parallels them. Given the published accuracy of ArcUSA, a question which occurs is: Where is the interstate more than 1792 meters away from the GPS track? You can find this out by making the point-based GPS track active and then selecting those points of the GPS track which are within 1792 meters of the selected parts of Rds2m, which are I-75.

{__} Make I090317A active. (Leave I-75 selected; it should still appear in yellow.) Pick "Select By Theme" from the "Theme" menu. You want to select the features of the active theme that are within a distance of 1.792 kilometers of the selected features of Rds2m. Make a new set of these features.

{__} The dots that remain red are those which are not selected, and are therefore outside the 1792 meter average accuracy distance.

{__} Zoom to the GPS track extent. Get the GPS track table. Note that 559 of 682 points were selected. "Switch Selected" and get 123. What percentage of the points were outside the active part?

_____

# Part 7

# The Present and the Future

IN WHICH *we examine several facets of GPS and GIS that are of current and future interest.*

# The Present and the Future

## OVERVIEW

By this time you have a good grasp of many of the fundamentals of GPS and, if you have done the projects, some facility in combining GPS and GIS. I now want to introduce you to some additional topics — not simply refinements of what you learned before, but truly different facets of the GPS phenomenon. The subjects presented will be:

- obtaining GIS feature attribute data with GPS equipment,
- navigating with GPS,
- real-time differential GPS position finding, and
- methods of planning for important GPS missions.

Finally I will present some discussion of trends in GPS and GIS use that should at least hint at what will take place in the remainder of this century. After that, it's anybody's guess.

## Obtaining GIS Attribute Data with GPS Equipment

From a software point of view, a GIS could be defined as the marriage of a graphic (or geographic) database (GDB) with other databases — most frequently a relational database (RDB). These other databases — which contain **attribute** data about features in the GDB — are usually textual in nature, but could consist of images or sounds. (For example, you could key in a street address and be shown a photo of the house there.) The combination of a GDB and RDB allows the user to make textual queries and get graphical responses (e.g., show with a red "X" those streetlights which have not been serviced since August 1995) or, conversely, make a graphical query and get a response in text (e.g., indicate the daily yield from parking meters in this area that I have outlined using a mouse on this image of the city).

If a GIS is a database with attribute information about geographical features, then it seems reasonable to collect the attribute data at the same time the positional data are collected. Thus far in this text we have not done this with GPS. The process of adding attribute data occurred after we generated the GIS coverages, since our GPS files contained only fixes in 3D space. Suppose you collected a sequence of fixes along a two-lane road. Once the arc was depicted in the GIS, you would have to add the "two-lane" fact to the record which related to the arc, if you wanted that information in the database.

Probably the most efficient and accurate way to use GPS to develop a GIS database is to collect the position data and the attribute data at the same time. Since a human operator is required to take the position data with a GPS receiver, it makes sense to have her or him enter the attribute data as well. Some GPS receivers allow this sort of data collection.

## The Organization of Attribute Data

The entry of attribute information into a GIS by using a GPS receiver is facilitated by a **data dictionary,** which is a hierarchical collection of textual terms stored in the GPS receiver's memory. The terms fall into three categories:

- Feature
- Attribute
- Value

**Feature** is used as it is in ARC/INFO. It refers to the feature-type that is the subject of a coverage; it is the "Feature" of "Feature Attribute Table" such as a "PAT" (Point Attribute Table) or "AAT" (Arc Attribute Table). Examples of feature-types are parking meters, paved roads, and land use. A point (or the average of a set of points) collected by the receiver under a given feature may become the basis for a record in a PAT. Or a sequence of points recorded along an arc may become the basis for a record in an AAT.

**Attribute** is also a concept parallel to one in ARC/INFO. Attributes are the "items" or "columns" of a feature attribute table. Suppose, for example, you are developing a GIS database about parking meters. In addition to the positional data collected at the meters automatically, you might wish to be able to record the meter identifiers as one attribute, and meter conditions as a second attribute.

**Value** refers to the actual entries in the table. Continuing our parking meter example, you could enter the meter number as the identifier of the parking meter and select a menu item "fair" to indicate the condition of the unit.

As you can see, there is a hierarchy to these terms: Features contain attributes; attributes are columns of values.

Once you have collected feature data with a GPS receiver, if you are converting these data to ARC/INFO coverages, each feature-type[1] becomes a separate ARC/INFO coverage with its own feature attribute table. The table consists of the usual initial items (area, perimeter, internal, and user_ids for a PAT; from and to nodes, left and right polygons, length, and user_ids for an AAT) plus columns for each attribute for which there is data. If a given feature (record) has a data value for a given attribute, that value

---

[1] The word "feature" has at least two meanings. It can mean the subject of a coverage (e.g., historical markers) or it can mean a particular entity (e.g., historical marker number 876). So I will use "feature-type" to indicate a general class of features, and "feature" to mean a particular entity.

becomes an entry in the feature attribute table.

## The Data Dictionary

A data dictionary need not be long or complicated. For example, consider the following one, named Very_Simple.

```
Rocks (point)
Trees (point)
```

You could make this data dictionary on your PC using the PFINDER software. Then you could transfer it to your GeoExplorer receiver.

To collect feature data with the Very_Simple data dictionary, you might begin a GPS file (for example, A010101A), asking the receiver to record a point every ten seconds. You might then move around the area of interest. When you arrived at a tree, you would select the feature "Tree" from the GeoExplorer menu and collect a number of fixes. You would then close the feature and, perhaps, move to the site of another tree, opening the feature, collecting data, and closing the feature. Should you encounter a rock you could open the feature "Rock" and record fixes there. The fixes you record at each individual object would be averaged to produce a single point which approximated the position of the object.

Upon returning to your PC, you would use PFINDER to transfer the file A010101A, and generate A010101A.SSF.

Then you could use A010101A.SSF to generate the appropriate ESRI files. From these you could, of course, make the coverage A010101A, which would consist of all the fixes collected. But by specifying that you wanted feature output, you could make two additional coverages, called something like:

- ROCKS1
- TREES2

ROCKS1 would contain a single point for each rock you visited. The point for a given rock would be the average of the fixes collected at that rock. That point would, therefore, provide approximate coordinates for each rock. A PAT would be formed with several records, one record for each rock. (The same idea would hold true for TREES2.)

You might instead use a somewhat more complex data dictionary, Still_Simple, that might look something like this:

```
Rocks (point)
   Size (menu, no default value)
      Small
      Medium
      Large
   Color (menu)
      Black
      Gray (default value)
      White
Trees (point)
   Size (numeric field, between 2 and 300,
         default value 10, two decimal places)
   Type (character field, maximum length 10)
Streets (line)
   Width (number of lanes, integer numeric field)
   Pavement (menu)
      Blacktop (default value)
      Concrete
      None
Intersections (point)
   T_intersection (menu, no default)
      Yes
      No
```

Assuming data were taken for all the feature types (Rocks, Trees, Streets, and Intersections) five coverages could result: A010101A, ROCKS1, TREES2, STREETS3, and INTERSE4 (coverage names can be a maximum of 8 characters long).

In addition to the standard items (columns, fields) in the PAT, the coverage ROCKS would contain the items SIZE and COLOR. The values which could appear in these columns would be "Small", "Medium", and "Large" or no value at all for SIZE. What values could appear in the COLOR item? _____, _____, _____.

## From the Environment, through GPS, to GIS

The process of recording attribute data with a GPS receiver is a good bit more complex than simply recording position data, which is itself, as you know, not a trivial matter. To record attribute data you have to go through several steps:

- Build a Data Dictionary with a computer. This can be done on a PC using the PFINDER software. (Unfortunately, PFINDER is needed both to build the data dictionary and to transfer it to the GeoExplorer. If you only have access to GEO-PC you cannot collect attribute data.)
- Load the data dictionary file into the GeoExplorer. This process is similar to transferring position files and almanacs from the receiver to the PC; the data simply go the other direction. The PC may contain a number of data dictionaries; the GeoExplorer stores only one at a time.
- Take the receiver to the field. Select a particular feature-type from the menu. While the unit is automatically collecting position information, you manually select the appropriate attribute and value items. When enough fixes have been collected, stop the data collection process for that feature (i.e., close the feature). The fixes obtained will be averaged so that a single point represents the given point feature.
- Continue to collect data in the field. The data collected when no feature is selected may ultimately become one ARC/INFO coverage. For each feature-type for which you collect data, an ARC/INFO coverage may be built. That coverage will contain the number of individual features for which you recorded position information. Note that a given data collection session could result in any number of ARC/INFO coverages, the data for which might all be contained within one GPS file.
- Close the GPS file. Open a new one if you like and collect feature data, or simply positional data, as you wish.
- Return to your office or lab and upload the files from your GPS receiver to the PC in the usual way.
- Using PFINDER, produce the SMLs and associated ESRI files that, when executed, will produce ARC/INFO coverages.

To practice using the feature attribute collection abilities of GPS, do PROJECTS 7A, 7B, and 7C.

## Navigation with GPS Equipment

GPS was not developed primarily to allow users of GIS to obtain better data. Rather, the purpose, which we have virtually ignored in this text, was to provide better navigation and instantaneous position information. You noted, back in Part 1, that you could determine your direction and speed of travel as you were walking. The software built into the GeoExplorer allows you to perform a number of other navigation functions.

A **waypoint (WPT)** is a single, 3D position that can be stored in the memory of the receiver as a special point with a number and a name. A waypoint is a beginning point, a destination, or simply a "point on the way" to somewhere. The GeoExplorer receiver can store up to 99 waypoints.

A waypoint can be set up in the receiver by:

- manually entering the coordinates of a point, or
- copying the current position of the receiver.

Obviously, before you enter any waypoints manually you must be certain that the datum, coordinate system, and altitude reference are set properly in the receiver.

Once a waypoint is stored, you can navigate to it by reading various screens on the receiver.

To practice using the navigation abilities of GPS, do PROJECT 7D.

## Real-Time, Differential GPS Position Finding

You learned about differential correction in Part 4. We discussed post-processing correction in some detail. While you can get much more accurate fixes from such a system, it requires a number of additional steps which are time consuming. Once you know the source of the differential corrections file, you have to get it, load it onto your computer, and execute software to produce the final corrected file. Wouldn't it be nice if the GPS unit simply gave accurate positions at the time you took the data?

Further, there are some applications in which there simply isn't time to post-process GPS fixes — bringing a ship into a narrow harbor, for one example, or landing an airplane when the pilots cannot see the ground, for another.

Well, actually there is a way to provide instantaneous, accurate fixes: Real-Time Differential GPS.

## Getting Corrections for GPS Measurements: Right Now!

**Real-Time Differential GPS (RDGPS)** may be approached in several ways:

- The user can set up a base station over a known point and arrange for transmission of a radio signal from the base station to roving receivers.
- The user's GPS setup can receive correction signals broadcast from an antenna in the area operated by a corporation or a government agency. Such installations operate a base station continuously and broadcast the correction data, sometimes as part of another signal such as a standard FM broadcast. The U.S. Coast Guard maintains such a service in some coastal areas, and companies such as Accupoint, Inc. and (Differential Corrections, Inc.) sell a correction service.
- The user's GPS setup can receive correction signals from a communications satellite parked over the equator. These signals come from data taken by base stations located in disparate parts of the United States. The data are analyzed, packaged, and sent to a geosynchronous communication satellite for rebroadcast to Earth. The system uses this approach with a satellite located on a meridian which passes through Lake Michigan.

With all three of these methods, many of the requirements for base stations remain the same. They must take data from all satellites that the rover might use. They must take data frequently (every few seconds, at least). They should have a separate channel for each satellite so as to track it continuously.

With all the methods of doing real-time differential correction, you might wonder how these pieces of equipment manage to talk to each other. While the frequencies on which broadcasts take place may vary, the content of real-time differential correction data is standardized. The current standard is RTCM SC-104 (version 2), promulgated by the Radio Technical Commission for Maritime Services, in Washington, D.C.

## A User-Operated Real-Time Base Station

The user may operate her or his own base station. The complications go up with differential correction (real-time or otherwise), of course. Here the user has to have at least two GPS receivers, plus a transmitter associated with one of them (the base station) and receivers for each rover. In addition to the roving GPS receiver and all of its settings and conditions for good position finding, the user has to also be concerned about the same factors for the base station, plus making the radio link between the stations work at the distances required. Additional complications relate to mounting the base station antenna at a precisely known point. Those who need really precise coordinates relative to some nearby known point (e.g., land surveyors) use this method frequently.

## A Centrally Located Real-Time Base Station

Since GPS signal errors tend to be quite similar over wide geographic areas, there are obvious advantages to having a single base station serve for any roving stations in that area. Put another way, there are very few reasons for each of twenty-eight GPS users, who are in reasonably close proximity to each other, to collect and rebroadcast identical correction data. This obvious fact, plus the entrepreneurial nature of American society, has produced GPS differential correction services, wherein a user contracts for equipment and the right to receive corrections to the raw GPS signal. Users who are sufficiently close to a U.S. Coast Guard GPS beacon may pick up a signal at no cost besides that of the receiving equipment.

## A "Differential Corrections Anywhere" System

The problem with "centrally located" base stations is that there may not be a signal from one where you are trying to take data. While there are more and more base stations every month, vast gaps exist where their signals do not go. A solution to this problem is a system in which the broadcast of the correction data comes from a communications satellite. This means that, for a large portion of the Earth's surface, the user's receiver is never out of range in the area covered by the satellite, though local topography may block the signal. The satellite of the only system that currently employs this method (OMNISTAR) can be as low as 30 degrees above the horizon in northern parts of the U.S. The satellite broadcasts a straight line signal of about 4000 megahertz. The area covered is all of the U.S. (except parts of Alaska) and parts of Canada and Mexico.

OMNISTAR has eleven base stations located on the periphery of the U.S. These stations transmit data regarding the errors in the GPS signals in their areas to a central network control center. These data are then analyzed and repackaged for transmission to the communications satellite.

The broadcast from the satellite is such that the data from the eleven stations can be tailored to the user's position. How does the OMNISTAR system know where the user is? The user's GPS unit tells the attached OMNISTAR radio, which then decodes the signal from the communications satellite to provide the proper corrections for the local area. Since a real-time differential correction datum is useless unless it arrives at the GPS receiver within moments of the time a fix is taken by the receiver, you can see that a lot has to happen in an extremely short amount of time. (A GPS signal code leaves a GPS satellite; it is received by your receiver and by the OMNISTAR base stations; the base stations transmit the signal to a central location where it is processed and sent to the communications satellite, where it is resent to the radio next to your GPS receiver; that radio determines the correction your GPS receiver needs and supplies it. Whew.)

This system can provide measurements such that 95% of the fixes taken lie within half a meter of the true point and the mean value of a number of points is accurate to within a centimeter (horizontally), using a top-of-the-line survey grade receiver. With a GeoExplorer, you can expect 95% of the fixes to lie within nine meters, with means within two meters of the true point.

To practice using real-time DGPS, do PROJECT 7E.

## Planning the GPS Data Collection Session

Fact: The GPS data collection process in a given area is better at some times than at others. Determining a good (or just satisfactory) time to take data is called **mission planning.**

Not too long ago, the discussion of mission planning would have had to appear at the front of the book, because there were so few GPS satellites in orbit that you had to go into the field at specified times to collect data. Now with the NAVSTAR system at full operational capability, you can almost always "see" enough satellites to get some sort of fix. But is it a fix that is accurate enough to meet your needs? (Our needs in this book were simply to show you how the system worked. So we left the mission planning details until now.)

It would doubtless have been simpler if the 24 GPS satellites were simply parked over various areas of the world so that, in a given location, you always dealt with the same set of satellites in the same positions. Unfortunately, this would mean that they were always parked over the equator, as, due to the laws of physics, all geosynchronous satellites are. If they were orbiting the Earth anywhere else, as they would have to be in order to provide world-wide coverage, they would be moving with respect to the ground underneath.

So the rule for GPS satellites is "constant change." The number of SVs your receiver can see changes. The geometric pattern they make in the sky changes. If you want to collect data at the optimum times — for examples, when the DOP is low, or when there are a large number of satellites available for overdetermined position finding — you might want to use mission planning software. This software will also help you if your data collection effort might be impeded because there are barriers between you and some part of the sky — you would like to have information about the expected behavior of the satellites. You may use the software to simulate such barriers and to tell you when the satellite configuration will be such that you can take good data anyway.

## Almanacs

A **GPS almanac** is a file that gives the approximate location of each satellite. Actual position finding does not depend on the almanac; a much more precise description of each satellite's position is needed for a GPS fix calculation. (Precise descriptions of the orbit come from an ephemeris message, broadcast hourly by the satellite itself.) Rather, the almanac is part of the general message which comes from each satellite, describing the orbits of all satellites. The almanac gives parameters from which the approximate position of all satellites can be calculated at any time "t" in the future. Since it doesn't have to be very precise, an almanac may be good for months; left to themselves, GPS satellites deviate little from their projected orbits. However, the DoD may move a satellite forward or back within its orbit — perhaps to replace a failing satellite or to make room for the launch of a new one. Activities like this make it necessary to collect almanac information frequently. In fact, the GeoExplorer collects an almanac every time it is outside and tracking one or more satellites for a minimum of about fifteen minutes. This almanac can be transferred to your PC. While the receiver will not tell you the almanac's age, the GEO-PC software will tell you if an almanac is more than a month or so old. An almanac may also be loaded onto a PC across a network from an electronic "bulletin board" maintained by the manufacturer of your GPS equipment.
    Almanacs are used for at least two purposes:

- giving information to the receiver so that it can locate the satellites more quickly, and
- providing information to PC software so it can tell the user the best times to collect data.

It is this second purpose we examine here: using a PC to tell us when we can get the best data at a given location.

## Mission Planning Software

Several software aids exist to help a GPS user determine the best time to collect data. Two that we will look at are:

- the "Plan" module found on the GEO-PC main menu, and
- a Windows program called Quick Plan.

To practice using the mission planning software, do PROJECT 7F and, if you have Quick Plan, PROJECT 7G as well.

## On the Horizon: Trends

I read recently in a GPS trade publication something to the effect that GPS receivers are becoming so integrated with other equipment that GPS itself is becoming virtually invisible. "Oh, great!," I thought. "Just when I've spent all this time and effort learning about it." You may have a similar reaction.

Upon reflection, I realized that the observation, while true for many uses of GPS, probably will not apply to GIS applications. Most users of GPS make immediate use of the GPS signals and do not record them for future processing. The use GIS professionals make of GPS development of datasets whose quality must be high and which will endure for months, years, or even decades. But it is certainly useful to look at the trends in GPS use across the board, to more fully understand this amazing phenomenon. The following seem to me to be fairly safe predictions:

- Accuracy will increase, both gradually and by significant steps.
- The time required to obtain a fix of a given accuracy will decrease.
- GPS will be combined with other systems to provide positional information in places of poor or limited GPS reception.
- "Monumentation" — the activity of providing markers on the Earth to indicate exact geographic location for reference by those who need such information — will increase and change radically in approach.
- Air navigation will be radically transformed.
- Ground vehicle navigation will result in more efficient use of resources.
- Reliability will increase dramatically for some applications.
- New satellite navigation systems will emerge, perhaps complementing NAVSTAR, perhaps competing with it.
- GPS signals for some applications will originate from pseudo-satellites on the ground.
- Civilian and military sectors will increase cooperation; the civilian side of GPS becoming relatively more and more important.
- GPS will be the principle mechanism for distribution of accurate information about time.
- Emphasis will shift from using GPS to provide information to humans who control processes and artifacts, to direct control of those processes and artifacts by GPS.
- More applications of GPS will emerge; those in use will become more extensive.

A brief discussion of some of these trends follows.

## Better Accuracy

In addition to the gradual increases in accuracy brought about by improved equipment and processing methods, a couple of factors may dramatically improve what we might expect of GPS in the future.

Losing "selective availability" will be a wonderful start. A committee of the National Research Council, commissioned by the National Science Foundation, unanimously recommended that the DoD's Air Force Space Command discontinue the distortion of the signals. The DoD NAVSTAR managers have agreed, apparently encouraged by the U.S. civilian administration (Vice President Gore made the announcement in March of 1996) but the lack of alacrity is obvious: phasing out could take a decade.

The wide availability of DGPS signals will increase and improve, as will the types of origins of such signals. Between the Coast Guard Beacon network, commercial firms offering FM subcarrier signals, and OMNISTAR, DGPS is pretty much available in the U.S. If SA goes away, is this all wasted effort? No. For really precise work, DGPS is required anyway.

Military receivers are able to be more precise than civilian ones (even with SA turned off) for a variety of reasons. One is that they receive on two frequencies. Comparison of the two signals allows a better estimation of the error caused by the ionosphere. Another civilian frequency is being discussed — but is not likely soon.

## Faster Fixes

While GPS at present can provide surveyors with the centimeter-level accuracy they require, the length of time they need to "occupy" a location is usually prohibitive for routine land surveying activity. The kind of accuracy surveyors need requires that the receiver stay in constant touch with the satellites, so that they may actually count the number of wave cycles of the radio signal between the antennas. This is termed **"carrier-based"** position finding, as differentiated from the GPS work that you have studied. In areas or conditions where the receiver may "lose lock" on the satellite, the process can be not only long, but frustrating. Better equipment and the availability of carrier phase DGPS signals will begin to make GPS viable for many surveying applications.

## GPS Combined with Other Systems

It has been commented that GPS bears a resemblance to sanitation plumbing of a century ago: basically an outdoor activity. But often we want to know where we are when we don't have the luxury of a line-of-sight relationship with four or more satellites. In tunnels or mines or caves, for some examples — even in deep valleys or on urban streets — GPS operates at a disadvantage. One way to locate your position, assuming you knew it at some time in the past, is an **Inertial Navigation System (INS).** An "INS" is a mechanical apparatus, whose heart is a set of spinning gyroscopes; it can detect very small changes in its position and report these to an operator. Marine navigators can place position data into such systems while still in harbor, sail for three months, and return to the point of origin, with the "INS" indicating the original settings within a hundred yards or so. Certainly it is possible for a vehicle which is mapping a road system to have an INS on board to see it through tunnels, or valleys between skyscrapers.

The Ohio State University Center for Mapping has developed a van for delineating highways that combines a number of position locating techniques and systems with photographic or videographic information, showing pavement condition, signs, intersections and the like. The information can be installed in a GIS. Later, a point on an interactive computer map can be selected and the user can "drive" along the road, seeing the features.

## Monuments Will be Different, and in Different Places

Things are changing at the National Geodetic Survey agency office — more in the past ten years than in the preceding two centuries (well, almost; Thomas Jefferson set up the Survey in 1807). Monuments used to be on the tops of mountains, since line of sight was vital to the surveying profession. Line of sight is still vital, and still in the upward direction, but towards the satellites. And instead of static monuments we can expect radio-dynamic ones, that announce where they are. Further, the physical monuments will be more accessible and more closely spaced — and a lot more accurate, having been "surveyed-in" with GPS techniques.

The charted positions of many places have been changed based on GPS information. Numerous islands have been moved from their originally charted positions. The coordinates of the city of Hannover, in Germany, shifted a couple hundred meters to the south as worldwide systems replaced national ones.

## Air Navigation Will Be Radically Transformed

The U.S. Federal Aviation Administration recently awarded a contract for the Wide Area Augmentation System (WAAS) for $475 million, whose primary basis is GPS. Ground-based navigation stations and towers will become obsolete. So will ground-based radar units. Today, to keep track of a flight, a flight controller's radar sends a pulse which strikes the airplane and triggers a "transponding" pulse from a radio carried in the aircraft. This gives the controllers on the ground the 2D information they need to tell the pilot where to fly to avoid other aircraft. But the altitude information has to come from the airplane's own altimeter, whose readings are based on barometric pressure. With GPS, the airplane's electronics will simply know where it is, in all three dimensions, and that information will be sent directly to the controller.

That is, if there is a controller. It may be that an intelligent, computer- and GPS-based system will provide navigation and collision-avoidance information directly to the pilot.

That is, if there is a pilot. Since the GPS-based system could certainly control the aircraft . . . no, no, I take this too far. Surely there will always be pilots.

One concept that is being discussed is doing away with many of the airways. At present, aircraft operating under instrument flight rules (IFR, where pilots whose planes are in the clouds avoid hitting the ground by reference to charts and instruments, and avoid hitting other planes by being told where those planes are by a radar controller on the ground) follow highways in the sky. And just like highways on the ground, they do not usually lead directly from the trip's origin to its desired destination. One reason for the detours is collision avoidance. A second is that navigating the airplane is based on having to go through intersections (where the limited-range, ground-based antennas are). The combination of GPS and "real-time GIS," which shows the location of other aircraft, can obviate both of these factors.

If every plane could fly "direct," detouring only when the GPS/GIS system suggested that the current course might result in bent metal, vast amounts of time and resources would be saved.

In addition to navigation en route, GPS systems, augmented by real-time differential techniques, have shown abilities to provide positional information of less than half a meter to landing aircraft.

## Marine and Vehicle Navigation Will Be Improved

Something on the order of a million in-car GPS navigation systems are being installed each year. Not only do these systems allow navigation based on the static facts of where the roads go but also on the dynamic facts relating to where the other traffic is. Tokyo will soon benefit from such a system.

In addition to vehicle route guidance, intelligent transportation systems will provide other benefits, such as fleet vehicle dispatch and tracking, and emergency notification of police and tow trucks. Before you cheer too loudly, however, realize that without proper safeguards, Big Brother may know where you are, where you are going, and how fast. Already, automated systems exist in the U.S. that scan license plates of speeding cars and send a ticket to the owner.

## System Integrity and Reliability — Great Improvements

While you might think that the maximum safeguards would be provided for a system designed to aid a warrior with a mortar or a nuclear-tipped ICBM, the fact is that GPS position data can suddenly disappear for periods of a few seconds to a few minutes. That state of affairs is not going to be allowed for some important civilian applications. You will understand the need for reliability immediately if you imagine yourself as a passenger in a landing commercial aircraft. The plane is fifty feet above a runway in a blinding snowstorm. That's no time for GPS to take a breather.

It is going to be possible for ships to navigate in shallow water by knowing their vertical height and having an integrated survey of the bottom of the water body. On-the-fly marine navigation requires 3D GPS decimeter (10 centimeter, 4 inch) accuracy. Again, everything has to work right and keep working.

GPS will be vital to all sorts of civilian and commercial endeavors, so the systems built around it simply must be highly accurate and reliable.

## Other Countries, Other Systems

Since the U.S. is very proprietary with the more accurate aspects of the NAVSTAR GPS, people in other parts of the world look with concern at hanging their navigation and position recording capability on a system over which they have no control.

Of course, there is already another GPS system up and running: GLONASS has almost the same number of operating satellites, and has been around about as long. (The first GLONASS satellite was launched in October 1982.) And the Russians do not dither the GPS signals.

The European community is planning its own GPS (or maybe "EPS," since it will primarily cover Europe): the Global Navigation Satellite System (GNSS), which may or may not include GLONASS. They also plan on their own system for aircraft navigation, based on satellites parked over the equator: the European Geostationary Navigation Overlay Service (EGNOS).

The Japanese also want self-sufficiency in something as important as satellite navigation systems. Their interest has been sharpened by the number of large earthquakes they experienced in 1995. Can GPS aid in predicting earthquakes? They certainly think so. And they have big plans for use of satellite positioning systems in aircraft and automotive transportation: the Multi-function Transport Satellite Service (MTSAT), which is a geostationary satellite overlay service like WAAS and EGNOS.

## Civilian and Military Interests Will Cooperate

Once the civilians and the military agree on a few more issues related to GPS accuracy, coverage, and integrity, GPS will be the primary U.S. government radio-navigation system well into the next century. The satellites which will be launched to provide GPS information in 2015 A.D. and beyond are already being designed. Swords are being beaten into plowshares, but the changes will not come rapidly, for both political and security reasons.

A real breakthrough would be the development of a way to jam GPS signals in a given geographic area, thus denying an enemy any use of the civilian GPS signal. As it is, all are denied use of accurate signals, unless they use DGPS — which is becoming ubiquitous.

## GPS Will Become the Primary Way to Disseminate Time Information

A GPS receiver is one of the most accurate clocks in the world, *if* it has continual access to the satellites. The forte of a clock in a GPS receiver is short-term accuracy, not long-term consistency. People who really want to know exactly what time it is can set up a base station over a known point and analyze GPS signals for time instead of position. GPS time does differ from UTC time by an integer number of seconds, such as 10. There are many applications for which coordinated time is vital — making the internet, on which the World Wide Web is based, run smoothly, for example.

## GPS: Information Provider or Controller?

There will be an increasing number of applications in which GPS signals control equipment directly, rather than going through a human "middleman". In such joint GPS/GIS uses as fertilizer or pesticide application, the automated system may steer the tractor while the farmer rides along simply for reasons of safety. Carving out roadways or laying pavement may be conducted in similar fashion. The Center for Mapping at Ohio State University boasts a system that can put a bulldozer blade in the correct position with an accuracy approaching one centimeter.

## Applications: New and Continuing

New applications appear every day, from vital operations such as golf cart tracking and pin ranging, to incidental ones such as aiding farmers in feeding the hungry people of the world by enhanced agricultural techniques. Animals can be tracked by researchers. Polar ice sheets can be monitored. Blind people can be guided. Rescue helicopters can be directed. Solar eclipse tracks and weather information can aid navigating astronomers. Chase cars can follow hot air balloons, even when they can't be seen. And in space: some satellites carry GPS receivers! What better way to report on exactly where they are?

The magazine *GPS World* reports monthly on new applications, and sponsors an applications contest once a year. And, of course, the World Wide Web has information of interest to those searching for information on GPS — more than 3000 pages.

## GPS and GIS

The direct integration of GPS and portable computers running GIS is a reality, allowing new facility in navigation and data collection. The company GeoResearch developed GeoLink software which linked GPS input to a microcomputer running GIS software. Trimble Navigation has two products based on PCMCIA cards — which may be inserted into a socket in a notebook or "pen" computer — which integrate GPS signals, RDGPS correction broadcasts, and GIS. The "ASPEN Card" and "Direct GPS" turn a standard computer into a datalogger and navigator. Direct GPS displays position information, with ArcView providing the software context.

# STEP-BY-STEP

## Nature of the Projects

These projects relate to several of the subjects discussed previously in this Part. In these projects, which are somewhat advanced exercises, you will be expected to have and use the manuals which come with the equipment and software. The instructions contained herein are more general than those given previously.

Here is a "table of contents" for the various projects:

- Feature Attribute Data          7A,7B,7C
- Navigating                      7D
- Real-Time Differential GPS      7E
- Mission Planning                7F,7G

## Projects Utilizing Feature Attribute Data

{__} The following three projects (7A, 7B, 7C) relate to obtaining and converting GIS attribute information. If you have a GeoExplorer receiver and PFINDER software, and you want to collect data in the field, do all three projects.

If you have PFINDER, but do not want to take time to collect data, but do want to see most of the conversion process, do PROJECTs 7B and 7C.

If you do not have PFINDER (but do have ARC/INFO) do 7C.

{__} If you wish to do any of the projects, it is best if you make a separate directory to hold the data. Assuming you have done previous projects and have a "DATA_yi" directory, make a new directory under it.  Type the following from the DOS prompt:

**MD**   **\GPS2GIS\DATA_yi\FEAT_yi**

You will need this directory if you do any of the "attribute" projects. In any event, you should plan to read through all three projects.

## PROJECT 7A

### Obtaining GIS Attribute Information with GPS Equipment[2]

{__} In PFINDER, make a project based on the directory \GPS2GIS\DATA_yi\FEAT_yi. Make this the current project.

{__} *Make a data dictionary using PFINDER:*[3] Select "Utils", then "Data Dictionary". Pick "New". Use "Still_Simple" as the name. Use the form, if not the exact text, from the Part 7 Overview as a model for your data dictionary. (If you are in an urban area where rocks or trees are scarce, perhaps you could substitute POLES and HYDRANTS.)

A data dictionary is built with a cascading series of windows. You are prompted for feature-type name, attribute, and value, as appropriate. (Ignore blanks which ask for user codes.) "Okay" enters the particular entity; "Done" records your choices so far, then takes you back a level in the hierarchy. A "default value" is the one selected by the GeoExplorer if you don't select another. As you build the dictionary you will see it appear in a box on the right of the screen. You should be able to make "Still_Simple" with just the information found in the Overview section.

Once your dictionary is complete, experiment with the other menu items, such as Add Feature, Edit, Display, and List All.

---

[2] You must have PFINDER software to do the data dictionary building and actual data collection part of this project. However, you can do PROJECT 7C, in which you make ESRI coverages, using only PC ARC/INFO.

[3] You need the "green dongle" attached to the primary parallel port of your computer.

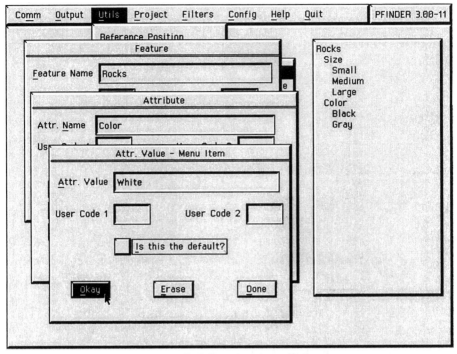

Fig. 7-1 — Building the data dictionary

{__} Connect the GeoExplorer receiver to the PC as usual, setting the proper communication parameters in both PFINDER and the receiver. Select "Data Transfer" from the main menu of the receiver.

{__} In PFINDER, pick "Comm". Transfer the almanac to the PC, just to be sure the connection is working.

{__} Transfer the "Still_Simple" data dictionary to the GeoExplorer:

{__} In the GeoExplorer, under "Main Menu ~ Data Capture ~ Dictionary ~ Review" examine the data dictionary you just loaded.

{__} Set up the receiver to collect position data. In addition to the parameters that you have set before, set the following under "Configuration ~ Rover Options":

- Set the "Not in Feature Rate" to 30 seconds.
- Make sure the "Hi Accuracy" setting is "off".
- Under "Feature Logging" set "Points" to 3 seconds.
- Under "Feature Logging" set "Line/Area" to 15 seconds.
- Under "Feature Logging" set the minimum number of positions ("Min Posn") to 10.

These parameters will hardly give you good accuracy, even if differentially corrected, but the object of this exercise is simply to have you practice collecting attribute data simultaneously with position data.

{__} *Collect some data:* Pick a mile or two long road segment that contains several intersections. Take the receiver out and open a rover file under the "Data Capture" menu Data should begin to be recorded in that file every 30 seconds.

{__} As you start your walk, go to "Select Feature". The "CMD" key here will give you a list of the possible features in the data dictionary, such as:

```
- Select Feature -
1.Rocks
2.Trees
3.Streets

4.Intersections
```

Pick "Streets". Data will now be recorded under that feature, so continue to walk. Note that the number of fixes recorded for the feature will be displayed in the upper right-hand corner of the screen. The following display will be presented.

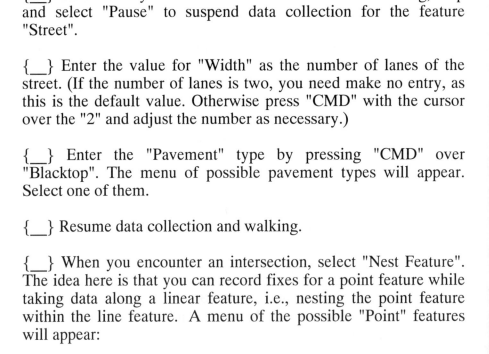

{__} Since it may be difficult to enter data while walking, stop and select "Pause" to suspend data collection for the feature "Street".

{__} Enter the value for "Width" as the number of lanes of the street. (If the number of lanes is two, you need make no entry, as this is the default value. Otherwise press "CMD" with the cursor over the "2" and adjust the number as necessary.)

{__} Enter the "Pavement" type by pressing "CMD" over "Blacktop". The menu of possible pavement types will appear. Select one of them.

{__} Resume data collection and walking.

{__} When you encounter an intersection, select "Nest Feature". The idea here is that you can record fixes for a point feature while taking data along a linear feature, i.e., nesting the point feature within the line feature. A menu of the possible "Point" features will appear:

```
┌─────────────────────────┐
│   - Nest Feature -      │
│    1.Rocks              │
│    2.Trees              │
│    3.Intersections      │
└─────────────────────────┘
```

{__} Select "Intersections". Fixes are now being recorded for that point feature, and you should begin entering the appropriate attribute data on a screen such as this:

```
┌─────────────────────────┐
│  Intersection    0      │
│  1.Close Feature        │
│  2.Pause                │
│  3.File Status          │
├─────────────────────────┤
│  4.Main Menu            │
│   T_intersection        │
│  Not Entered            │
└─────────────────────────┘
```

{__} The data dictionary requires you to enter a "Yes" or "No" as to whether the intersecting street terminates at the street you are on (a "T" intersection) or whether it goes through. Select "Not Entered", then choose the correct answer.

{__} Observe the number of fixes that have been collected for the intersection. If ten or more, close the feature. Data taking will resume on the line feature "Streets", so resume walking.

(To close a feature which doesn't have any fixes — or enough fixes — in it, put the cursor over "Close Feature" and press "CMD". If that doesn't work, put the cursor over "Close Feature" and press "Esc".)

If the GeoExplorer had a keyboard, entering text and numbers would be a lot easier. On the other hand, during a real data collection effort, you may have to occupy the point for a while so the limitation imposed by the receiver gives you something to do.

(As a math professor of my acquaintance remarked, regarding whether an automatic transmission or a manual transmission was preferable in an automobile: "Well, you might as well shift the gears yourself, since you have to be there anyway".)

{__} Nest other point features ("Rocks" and "Trees", or "Poles" and "Hydrants") that you find along the way, in addition to the intersections.

{__} At the end of your walk, close the feature "Streets".

{__} Close the file.

{__} As you walk back, open a new file, and the feature "Streets" within it. Close the feature and the file at the appropriate time.

{__} *Load the data into your PC:* Start the PFINDER software. Transfer the two files you just collected to the project containing the data dictionary: \GPS2GIS\DATA_yi\FEAT_yi.

{__} Differentially correct the files at this point if the base station files are available and you choose to.

## PROJECT 7B

This project is a continuation of PROJECT 7A, but initially uses some sample data provided in \GPS2GIS\FEATDEMO. I recommend that, even if you collected your own data in PROJECT 7A, you work through the exercises with the supplied data; then use the same process with your own data.

{__} In DOS, copy the file "F033119A.COR" from the FEATDEMO directory to your FEAT_yi directory:

```
CD \GPS2GIS
COPY    FEATDEMO\F033119A.COR    DATA_yi\FEAT_yi\*.*
```

{__} Start PFINDER and select the project that points to EAT_yi, where you developed the data dictionary.

{__} Under "Output" pick "Display". Prepare to display the F033119A.COR. On the "Display Options" screen put a check by "Display Positions" and "Join", using whatever color you wish, but using a simple dot for the symbol. Make certain that "Display features" is not checked. Okay. Observe the display. It contains all the fixes that were recorded.

{__} Set grid ticks to 100 meters.

{__} Add the same file again, but on the "Display Options" screen, "un-check" display positions and check "Display Features". Choose the "Line Color", "Point Color" and "Symbol" (something other than a simple dot — perhaps the square with a cross in it) from the box containing the "Display Features" option. Okay.

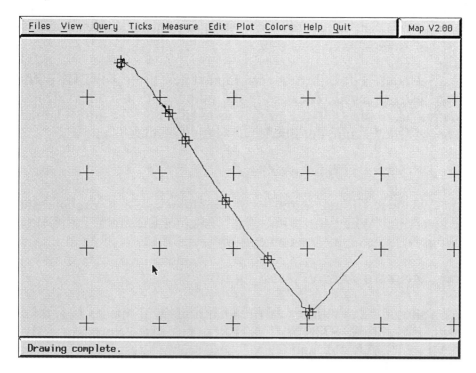

Fig. 7-2 — Display of "Street" and "Intersection" features

{__} Note that the line representing the "Streets" feature covers up most of the line representing the entire file. The fixes that are not covered by the fixes in "Streets" are those that were collected before "Streets" was started. Note also that single points represent the point features that were collected. The 10 or more fixes that were recorded at each nested point feature were averaged into a position.

{__} Now from the Main Menu select "Output", then "GIS". For the GIS pick "PC ARC/INFO — Tables".

{__} From "Options", select "Data Type". Select "Features Only". (If you selected GPS positions only, you would be preparing to make a coverage of the sort you generated in Part 5 — one with no attribute data.)

{__} Also under "Options", the number of vertices should be set to "All". Set the starting feature ID to 501. Under "Attribute Type" pick "Default". Make certain elevation data will not go into the output. Set the coordinate system, datum and units to UTM, NAD-27, and meters.

{__} Under "File", then under "Input Files", select the COR file you just displayed. Make sure the output path is correct. Direct output to a file (not the printer as well). For "Feature Output" check "One Feature Per File" and "Output Feature Name".

{__} Review the "Info" page.

{__} Under "Run", preview the graphics. Then execute the conversion. When the "time line" is complete, quit PFINDER. From DOS, examine the contents of your FEAT_yi directory:

```
CD  \GPS2GIS\DATA_yi\FEAT_yi DIR
```

You should see a number of file identifiers with names *beginning* with F033119A, STREET, INTERS, and extensions like SML, GML, PTS, AA, PA, and GEN. These are the files with which you can build ESRI coverages, which we do in PROJECT 7C.

## PROJECT 7C

### Creating and Viewing Coverages

A summary, mainly for those of you who did not have access to PFINDER: In PROJECT 7A we completed a data dictionary, installed it in the GeoExplorer, collected data under the structure of that dictionary, and brought those data back to the PC. In PROJECT 7B we looked at some similar sample data, and generated the SML and associated files necessary to make ESRI coverages.

From this point forward in the text, anyone with access to PC ARC/INFO may process the files which resulted from the previous work. If you did PROJECT 7B, skip ahead to the next section entitled "Executing SML Files". Otherwise, please copy and work with demonstration data I have supplied to create ESRI coverages. They are in \GPS2GIS\FEATDEMO:

- Files associated with STREET4
- Files associated with TREES3
- Files associated with INTERS5

{__} Use DOS to get these files into the FEAT_yi workspace you created at the beginning of this Step-by-Step section, using the following commands:

```
CD   \GPS2GIS
XCOPY  FEATDEMO\*.*    DATA_yi\FEAT_yi\*.* /S/V
CD   DATA_yi\FEAT_yi
```

### Executing SML Files

{__} Make certain that DOS is pointed at your FEAT_yi directory. Examine its contents. Note the names of the SML files. Type:

```
DIR   *.SML
```

{__} Go into ARC/INFO. Execute each SML file. For example,

**&RUN STREET4**

{__} LISTCOVS (or L -LC) to make sure the coverages were generated.

{__} *Investigate the coverages just created:* Start with the PATs and the AATs. For example,

**LIST STREET4.AAT**

You will get a single record, since the feature is a single arc. It is also a pretty dull record, with lots of zeros. The user_id should be "501", the value that was put in the starting feature ID.

Of considerable interest, however, is the additional attribute data. For the moment, skip over the item "Streets", which also has the value "Streets". There is the attribute item "Width" with the value "2", and the item "Pavement" with its value "Blacktop".

{__} Now list the PAT of INTERS5. There are several records. The user_ids range from 502 up — since the first feature taken, the arc, was numbered 501. In addition to the standard four items in a point attribute table, there are two additional attribute categories, one of which is "INTERSECTI" for which all the values in the table are "INTERSECTIONS". This came about because when the SSF and related files were generated, the box "Output Feature Name" was checked. (This also explains the item "Streets" and values under it ("Streets") in the STREET4 coverage.) If this identifying column were not included, the actual name of the feature — the name specified in the data dictionary — could be lost, since the coverage name is truncated and mutilated, and item names have fairly short maximum lengths. The maximum allowable lengths of item values are considerably longer. While this method of preserving the actual feature-type name may seem redundant and wasteful, it works, and another way is not readily obvious.

Examine the second additional item, "T_INTERSEC", for which the values are either "Yes" or "No".

In summary, then, separate, individual coverages are made for each feature type. The name of the coverage is an indication of the feature type name. Each individual feature recorded becomes a record in the feature attribute table. Each attribute is represented by a column in the table. And each attribute value becomes an entry in the table.

{__} Use ARCPLOT to examine the geographic image of the coverages:

```
ARCPLOT
DISPLAY 4
MAPEXTENT STREET4 INTERS5
ARCS STREET4 IDS
MARKERSYMBOL 97
POINTS INTERS5
POINTTEXT INTERS5 INTERS5_ID # LL
```

This should generate an image which looks something like this:

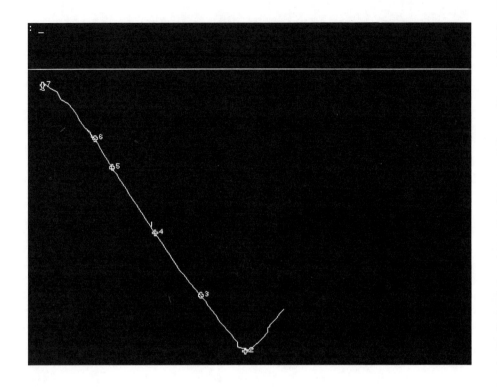

Fig. 7-3 — ARCPLOT of features collected with GPS

## Now Use Your Data

{__} If you did PROJECT 7A, go through the procedures of PROJECTs 7B and 7C, using your own data, using appropriate coordinates, datum, and units. (An important note: if you are using latitude and longitude as coordinates, make certain you have selected "Degrees" under "Angle" and a decimal accuracy of "6", all found under "Config" from the Main Menu. If you fail to do this, you will be attempting to feed ARC/INFO a coordinate format it cannot accept.)

## PROJECT 7D

### Navigating with GPS Equipment

In this project we will look at the various options for navigation.

{__} In the office or lab, set up your GeoExplorer: Under "Configuration ~ Units ~ Custom Setup" set the distance units to kilometers, the altitude reference to MSL, the north reference to true north, and the angular units to degrees.

{__} Under "Configuration ~ Coordinates" choose "Deg/Min/Sec".

{__} From "Position" on the main menu, determine the *longitude* and *altitude* of the "Old Position". Write these two values down.

{__} Go to the "Navigation" screen. It looks like this:

```
  - Navigation -
  1.Start Navigate
  2.To Waypoint
  3.From Waypoint

  4.Waypoint Setup
  5.Display Format
```

{__} Select "Waypoint Setup". The resulting screen appears:

```
┌─────────────────────────────┐
│  - Waypoint Setup -         │
│  1.Add Here                 │
│  2.Edit WPT                 │
│  3.Delete WPT               │
├─────────────────────────────┤
│  4.Clear All                │
└─────────────────────────────┘
```

{__} Select "Edit WPT". The up and down arrow keys change the number of the waypoint to be modified. Press the up arrow key until you find an unused waypoint (one with zero longitude and zero latitude). Press "CMD". Note the number of the waypoint (Wnn) that you are working with.

{__} *Give the waypoint the name "NP" for "north pole":* Use the up-down arrow keys to highlight the top line. Press the right arrow key to move the highlight to a single position to the right of the waypoint number. Press the up arrow until an "N" appears. Use the arrow keys to place a "P" next to it. (If you want to spell out "North Pole" and your thumb is up for it, go ahead. The lowercase letters are after the "Z".)

At this point you want to move on to the next field ("Lat"). If you press the down arrow key you merely get the next letter. If you press "CMD", you are asked if you want to save the waypoint. If you press "Esc" you are returned to the "Edit WPT" screen. So how do you move to the next line? This is where the previously unused "enter" key comes in. (It's the key at the upper right with the arrow on it.) Press it.

{__} Edit the "Latitude" using the same technique. Make it 90 degrees North.

{__} Edit the Longitude. Make it the same as your current longitude (as indicated by "Old Position", which you previously wrote down). Make sure you pick "East" or "West" correctly. (In virtually all the U.S. pick "W".) (What state is northernmost?

What state is westernmost? What state is easternmost?)[4]

{__} Edit the Altitude. Make it the same as the last position reading.

{__} Press "CMD". Save the waypoint.

{__} Find another blank waypoint. Edit it so that its name is "Eq" and that it represents a point on the equator at the longitude and altitude of "Old Position".

{__} Go to the "To Waypoint" screen. Press the up or down arrow keys until the "NP" waypoint appears. Press "CMD".

{__} Go to the "From Waypoint" screen. Press the up or down arrow key until "00" appears. Note that it is named "LAST POS" and has the coordinates of the last position recorded by the receiver. Press "CMD".

{__} Go to the "Display Format" screen which looks like this:

```
┌─────────────────────────┐
│ - Display Format -      │
│ 1.Dist/Bearing          │
│ 2.Track & XTE           │
│ 3.East/North            │
├─────────────────────────┤
│ 4.Velocity/Time         │
└─────────────────────────┘
```

Choose "Dist/Bearing". This will set the display so that distance and bearing are revealed during navigation.

{__} Although you are planning to stay in one position for the moment, begin the navigation process with "Start Navigate". You will see something like:

---

[4] Alaska. Alaska. Alaska.

```
┌─────────────────────────────┐
│  - Old Navigation  -        │
│  Dist:    5795.9Km          │
│  Turn:   left   118         │
│  Bearing:     0 Tn          │
├─────────────────────────────┤
│  Heading:    118 Tn         │
└─────────────────────────────┘
```

Most of the items on the screen won't make sense, because you are shielded from the satellites, and you are "proceeding" on the basis of "Old Navigation". But the distance should be the distance from "Old Pos" to the north pole and the bearing should show zero degrees. The heading, based on "true north" (Tn) and the amount you are to turn to get on course, should be the same. Write down the distance measurement: (OldPos_to_NP = _____ km.)

{__} Leave "Navigation" and configure the receiver to operate with a magnetic north reference. Then return and start the navigation again. The bearing now should reflect the **magnetic variation** (the difference between true north and magnetic north) in your part of the world. How much is it? _____

{__} Make "Eq" the "TO" waypoint. "Start Navigation". What is the distance from your position to the equator? _____
Add this distance to "OldPos_to_NP" which you wrote down earlier. The total should be about 10,000 kilometers.[5] What is it?
_____

{__} Change the display format to "East/North" and restart navigation. Here you are told how far to proceed in each of three dimensions to reach the destination waypoint.

_____

[5] The meter was originally defined as what was thought to be 1/10,000,000 of the length of the distance from the north pole, through Paris, to the equator. They missed by a little.

## More Walking

{__} Take the receiver outside so that it is calculating positions. Walk (or, if you have the remote antenna, drive) with it and experiment with the various display formats and the information presented.

{__} Set up a waypoint nearby and navigate to it.

## PROJECT 7E

### Real-Time, Differential GPS Position Finding

This project assumes that you have access to equipment that will permit real-time differential correction. Such equipment might be:

- A second GPS receiver, set up over a known point, transmitting RTCM form data that your GeoExplorer can receive if a separate radio is attached to it.
- A radio receiver connected to your GeoExplorer that is within range of a U.S. Coast Guard differential GPS beacon.
- A radio receiver connected to your GeoExplorer that is within range of one of the commercial DGPS service facilities (e.g., Accupoint or DCI).
- A C-band satellite antenna, receiver, and demodulator, connected to your GPS receiver (i.e., OMNISTAR).

Because of the wide variety of equipment configurations that can be used for differential correction, the following steps can only serve as indications of what is to be done.

{__} Before leaving your lab or office, make sure you have a good idea of what you are about to do. RDGPS is wonderful, in that it provides you with accurate fixes instantaneously. The downside is that you have a lot more equipment to cart around and connect. Make sure you understand and check out all the connections prior to taking the units into the field. Data flow between the radio[6] and the GPS receiver through the "download" cable you used when transferring files from the GPS receiver to a PC.

{__} *Find a site at which to take data.* If you can, find a site with the following characteristics:

- A known location, e.g., a survey marker.
- A clear view of the sky for receiving good GPS signals.
- Within range of a ground-based RDGPS signal, or, if you are using OMNISTAR, with a clear view of the sky to the south where the communications satellite is parked over the equator, due south of Lake Michigan.
- Where a source of power is available for the auxiliary radio receiver (possibly an automobile auxiliary power receptacle or an additional battery pack).

{__} Start acquiring GPS signals with the GPS unit.

{__} If you can, assure that the RDGPS radio is receiving signals from the transmitting unit, but do not yet connect it to the GPS receiver.

{__} Using the parameters for data collection provided earlier in this text, collect a file of about 20 fixes, using a time interval of 5 seconds.

{__} Connect the output cable from the radio to the "download" cable from the GeoExplorer.

{__} Under "Configuration ~ RTCM" you will find this menu:

---

[6] We will use the term "radio" to mean the unit which receives and processes the RTCM signals from the base station and passes the information on to your GPS receiver.

```
┌──────────────────────────┐
│        - RTCM -          │
│   1.Mode          GPS    │
│   2.Port            B    │
│   3.Stale time    010    │
├──────────────────────────┤
│   Msg Count         0    │
│   Msg Gaps          0    │
└──────────────────────────┘
```

{__} Choose "Mode". You will see these options:

```
┌──────────────────────────┐
│        - Mode -          │
│   Autonomous             │
│   Differential           │
│   Auto GPS-GPD           │
└──────────────────────────┘
```

"Autonomous" means the receiver calculates only uncorrected GPS positions. "Differential" forces the receiver to calculate, present, and/or file only positions corrected by RTCM signals. "Auto GPS-GPD" tells the receiver to calculate corrected positions when the RTCM signal is available, and to calculate uncorrected positions otherwise. (The uncorrected positions can be corrected later by post-processing if you wish.)

{__} Choose "Differential". This is also shown as **GPD** mode on some GeoExplorer screens.

{__} Under "Port" you probably want "A", but this depends on the requirements and cables related to the radio. Once you select the port you must select the communication parameters that relate to that port, by choosing options on a screen like this:

```
┌─────────────────────────────┐
│  - Comm Port A -            │
│  1.Protocol     XMDM        │
│  2.Baud         9600        │
│  3.Parity       None        │
│                             │
│  4.Data Bits       8        │
│  5.Stop Bits       1        │
│  6.Precision       2        │
│  5.Send At        2s        │
└─────────────────────────────┘
```

{__} *Change the Protocol:* The default protocol is XMDM, which is used when the GPS unit is communicating with a PC. You need to change this setting to the one appropriate for your radio. Probably this will be either "RTCM" or "RTCM/NMEA".

The GeoExplorer has the ability to *send* position and velocity data through the data cable in a format called "NMEA 0183".[7] You use this capability with a device that wants to know where your receiver is, such as the OMNISTAR radio.

(With most methods of real-time differential correction, the approximate location of your receiver is not a mystery, because it has to be within range of a signal generated locally and transmitted from an antenna on Earth. But the OMNISTAR signal comes from a satellite 22,000 miles away and can be used almost all over the U.S. (and in parts of Canada and Mexico). Since differential corrections vary from place to place, it is important that you receive the signals tailored for your area. Thus your receiver must tell the OMNISTAR radio where it is.)

See the RDGPS-receiver-operation manual for specifics on the protocols for sending data from the GPS unit to the radio.

{__} Set the baud, parity, data bits, and stop bits as required by your radio.

_____

[7] The standard for interfacing marine electronic devices.

{__} If you are using the RCTM/NMEA protocol (as you would if you were using OMNISTAR) you must set the NMEA precision, which indicates the number of decimal places in the latitude and longitude output. A value of "2" is suggested. (This provides for a precision to the nearest 0.01 degree. That increment of latitude amounts to about 0.7 miles — certainly close enough to choose a differential correction area. The rule of thumb is that each 60 miles on the ground generates about a 1 meter error in the reading.)

{__} Set the output rate ("Send At") of the NMEA signal. (Two seconds is adequate.)

{__} Back under "Configuration ~ RTCM" you will find "Stale Time". This is the duration that the GPS receiver will tolerate before giving up on receiving a RTCM signal with which to correct fixes. In GPD mode, data presentation and collection will stop if no RTCM message is received during this time. (In GPS/GPD mode, after the stale time has elapsed, the receiver will continue to calculate fixes, but they will not be corrected.)

{__} Observe the "Msg Count" and "Msg Gaps" indicators on this same screen. As you watch, you should see the message count increase by at least several per minute. The number of messages received as time passes is a good indication that you are receiving RTCM data from the radio. If this number does not increase, or resets to zero frequently, you probably are not getting satisfactory information from the radio. The problem could be in any number of places. Among them:

- the base station not transmitting,
- your RDGPS radio not receiving,
- bad cable connections, or
- communication parameters set wrong.

If things don't seem to be working, recheck your setup. If this doesn't work, ask your instructor.

{__} Check that your GPS receiver is still computing fixes. Look at the satellite tracking screen (under "GPS Status") and also be certain that the "Position" screen does not indicate "OLD".

{__} Assuming that everything is in order, collect 180 fixes in a file.

{__} Change the RTCM mode back to "Autonomous" and collect another file of about 20 fixes.

{__} Change the RTCM mode back to its default: "GPS/GPD".

{__} Change the protocol for the port you were using back to "XMDM" and the communication parameters back to those necessary to communicate with your PC.

{__} Back at your lab or office, transfer the three files you took to a PC. Examine them under "Display" with ticks set at 10 meters. The first and third files should show great variation, while the second file should show a tight group of points.


**PROJECT 7F**


**Planning the GPS Data Collection Session**

There are many ways to approach planning GPS missions; several software packages are available to help you. We'll look first at the "Plan" option in GEO-PC; then, for those who have access to the Quick Plan software (it usually accompanies PFINDER), we'll look at a more extensive program.

{__} *Get an almanac into your workspace:* A complete almanac file is recorded each time the GeoExplorer collects position data for more than 15 minutes. To check your unit for a complete file:

- turn on the GeoExplorer,
- let it complete its startup routine,
- get to the main menu,

- press and hold the on-off key, and
- observe the bottom line of the screen; you are looking for a *capital* "A", which indicates that the unit indeed has a complete almanac stored in its memory.

If, instead, you see a lowercase "a", the unit does not have a complete almanac and you will have to either take it outside to collect data for approximately 15 minutes, or find another source for an almanac. (Contact your GPS vendor for a current source such as an anonymous 'FTP" site, bulletin board, or address on the World Wide Web.)

{__} Connect the GeoExplorer to the PC as though you were going to transfer position data files. Turn it on. Select the "Data Transfer" option from the main menu.

{__} Start up the GEO-PC software. Choose the "Config" option from its main menu. Set the default path to your data directory: \GPS2GIS\DATA_yi\.

Choose the "Comm" option from the main menu. Pick "Almanac to PC". Click "Default". You should see a screen which gives the proper drive and directory. It also indicates a file_id of the form:

**yymmdd.ssf**

which represents today's date, *based on the date stored in the computer.* What is the name of the file? _____ This file_id actually has no connection with the date the almanac was collected. You could change its name to GEORGE.SSF if you wanted. The actual date of collection is, however, included in the almanac file itself. Okay the screen.

Shortly you will be told that the almanac has been successfully loaded.

{__} Select the "Plan" option. Choose "File". Make certain that the drive, directory, and file specification are correct for the almanac you just loaded.

{__} Pick "Setup". Under "Position" enter the approximate latitude and longitude of the location at which you want to collect position data. (Be sure you get east or west right.) In the label field write an identifier such as a town name.

{__} Under "Time Date" put in the year, month, and day you want to take data. (The default button brings in today's date.) Enter the correct time difference, to get results based on local time. Satellite positions and DOP change continuously, so the time spectrum must be subdivided into several reporting periods. "Interval" specifies the lengths of the time periods over which the computer summarizes the information it calculates and reports. Breaking the day into thirty minute periods is a good compromise between too much calculation and information, and too little.

{__} "Satellites" gives you the list of the PRN numbers of the healthy satellites, according to the almanac. Any unhealthy satellites are noted at the bottom of the screen. Probably you should just accept the list as presented.

{__} You may use "Masks" to set the PDOP values and the elevation mask you want. Setup is now complete.

The "Run" menu allows you to choose output of three sorts:

- Visibility. After some preliminary data on user location, masks, almanac name, and satellites considered, this output provides the times when 3D, 2D, or NO solutions are available. Totals are also shown.

```
Trimble Navigation LTD PFPLAN
  ver. 1.08 - GPS Visibility Information

User location:
  38 deg 02' N, 84 deg 36' W    Trimble Navigation Ltd.
Elevation mask: 15 deg  3D PDOP: 6  2D PDOP: 6     Interval: 30.0
Almanac: D:\GP_ALMAN.CUR\960310.SSF
24 Satellites Considered: 1 2 4-7 9 14-29 31
GPS Availability
 from Sun Mar 17 00:00:00 1996 to Sun Mar 17 23:59:00 1996 LCL
  LCL        Solutions Possible
00:30  3D solutions available
02:30  2D solutions available
03:00  3D solutions available

2D for 24:00        3D for 23:30
```

Fig. 7-4 — Satellite visibility report from GEO-PC

- Best PDOP. Indicates for each time the PDOP values and which satellites create the best geometry. Also shows all the satellites in view.

```
User location:
  38 deg 02' N, 84 deg 36' W    Trimble Navigation Ltd.
Elevation mask: 15 deg  3D PDOP: 6  2D PDOP: 6    Interval:
30.0 Almanac: D:\GP_ALMAN.CUR\960310.SSF
24 Satellites Considered: 1 2 4-7 9 14-29 31
Best PDOPs from Sun Mar 17 00:00:00 1996 to Sun Mar 17 23:59:00 1996 LCL

 LCL       3 Satellite          4 Satellite        Satellites Visible
                    (first 12 hours omitted)
12:00     19 22 31      1.76    18 19 22 31    4.47    18 19 22 28 29 31
12:30     19 22 29      1.79    19 22 28 29    3.83    18 19 22 28 29 31
13:00     27 28 29      1.75    18 27 28 29    3.67    18 19 27 28 29 31
13:30     27 28 29      1.71    27 28 29 31    2.93    18 19 27 28 29 31
14:00      2 15 28      1.61     2 15 19 28    2.27     2 15 18 19 27 28 31
                    (last 10 hours omitted)
```

Fig. 7-5 — Best PDOP report from GEO-PC

- Azimuth Elevation. Shows, at each time for each visible satellite (the SV number is the PRN number), the azimuth and elevation.

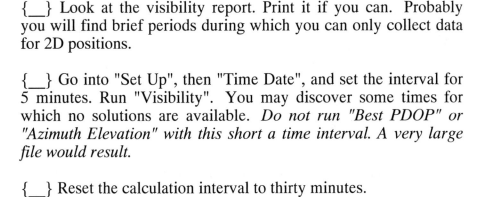

```
User location:
   38 deg 02' N, 84 deg 36' W    Trimble Navigation Ltd.
Elevation mask: 15 deg  3D PDOP: 6   2D PDOP: 6      Interval: 30.0
Almanac: D:\GP_ALMAN.CUR\960310.SSF
24 Satellites Considered: 1 2 4-7 9 14-29 31
Az El Table from Sun Mar 17 00:00:00 1996 to Sun Mar 17 23:59:00 1996 LCL

  LCL       SV  Az El   SV  Az El   SV  Az El  SV  Az El   SV  Az El   SV  Az El
                         (first 12 hours omitted)
12:00       14 205 03   18 284 57   19 300 28  22 052 31   28 096 56   29 197 60
            31 165 45
12:30       18 256 58   19 307 39   22 059 20  27 300 08   28 070 54   29 189 45
            31 155 59
13:00       02 258 03   18 232 52   19 312 51  22 066 11   27 306 18   28 054 45
            29 185 31   31 129 71
13:30       02 267 12   15 152 08   18 217 41  19 315 64   22 074 02   27 310 30
            28 046 34   29 182 18   31 079 71
14:00       02 277 20   15 144 19   18 208 29  19 306 78   27 311 42   28 043 21
            29 178 06   31 053 61
                         (last 10 hours omitted)
```

Fig. 7-6 — Report on satellite azimuth and elevation from GEO-PC

{__} Look at the visibility report. Print it if you can. Probably you will find brief periods during which you can only collect data for 2D positions.

{__} Go into "Set Up", then "Time Date", and set the interval for 5 minutes. Run "Visibility". You may discover some times for which no solutions are available. *Do not run "Best PDOP" or "Azimuth Elevation" with this short a time interval. A very large file would result.*

{__} Reset the calculation interval to thirty minutes.

{__} Run the best PDOP calculation. Note the variability in PDOPs. Note that you can tell approximately when various satellites rise and set (above the elevation mask). Note that a satellite is rarely visible for more than six hours; usually the visibility time is much less.

{__} Run the azimuth and elevation calculation. Print the result. Pick a satellite that was visible for a long time according to the "Best PDOP" printout and note its times of rise and set. Find this satellite on the "Azimuth Elevation" printout. Look at its positions each half hour; consider how it moves across the sky.

{__} Exit GEO-PC. At the DOS prompt obtain a directory of the files in \GPS2GIS\DATA_yi. Note that there are three files whose names are "CURRENT" and whose extensions are "VIZ", "DOP", and "AZL". These files were created when you ran the calculations for visibility, best PDOP, and azimuth-elevation, respectively. If you were not able to print out these files from GEO-PC, print them out with DOS, perhaps after transferring them to another computer.

## PROJECT 7G

### Mission Planning Using Quick Plan

Quick Plan is a sophisticated GPS planning program which runs under Microsoft Windows (version 3.1 and later). While the GEO-PC module "Plan" provides only textual output, Quick Plan exhibits some dramatic graphic representations of much of the same information. We will not delve deeply into Quick Plan, but you will see enough of it to let you know whether you want to learn more about it, which you can do from Trimble's *GPSurvey Software manual.*

{__} Bring up Microsoft Windows. From the Program Manager bring up "Quick Plan". From the "Quick Plan" window, bring up "Quick Plan".

{__} Select the same date as you used with GEO-PC — so you can later compare the output of the two programs.

{__} In the "Edit Point" submenu, select the same location as when you ran GEO-PC. Click on "Keyboard" to make any changes. Use the mouse to position the cursor to type onto the resulting screen.

{__} From the Quick Plan main menu, choose "Options", and then "Almanac". The file you want to load is the same one you used with GEO-PC. It is located in \GPS2GIS\DATA_yi\. From your experiences with Windows in general, or ArcView in Part 6, you should be able to navigate to this file. Ask your instructor if you need help. When you okay this screen, the almanac will be loaded.

{__} From "Options" pick "Time Zone". From the zone name dropdown menu, find your zone (remember about daylight savings time) and select it.

{__} Under "Session" pick "New Session". An "Add New Session" window will come up. The default "name" will be a four-digit number, the first three of which will be the Julian date, numbering the days from 1 to 365 (or 366), while the fourth is a sequence digit. You may change the name.

{__} The single location you specified a moment before will appear on the right of the window. Select it with a click. Then press "<Add<" to add the point to the empty list on the left side of the window. Okay the screen.[8]

{ } If a window such as that in the next figure is not on the screen, select "Options" from the main menu, followed by clicking on "Show Status." Examine the status window.

_____

[8] Quick Plan is a subset of a larger planning program for surveyors in which several sessions can be developed and saved even after the program terminates. This partly explains why there is a list of "all points" which contains only one — and why you have to transfer it to the current session.

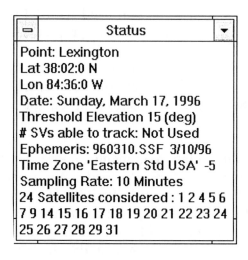

Fig. 7-7 — QuickPlan Status screen

{___} Select "Graphs". Pick "Number SVs and PDOP." A graph with two parts will appear. Enlarge the window with the up arrow in the upper right-hand corner. The graph should now appear something like:

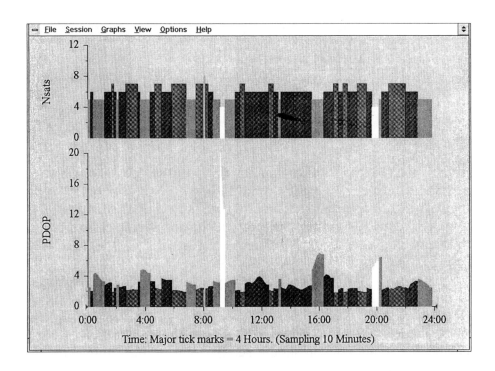

Fig. 7-8 — Graph of best PDOP and satellites visible from Quick Plan

On the top you see the number of satellites visible at each given time. The bottom shows the corresponding PDOP.

{__} From this window you can probably get a pretty good idea of when, and when not, to collect fixes. Avoid times of low satellite numbers and high PDOPs. (Leave this and subsequent windows open.)

{__} From "Options" set the "SV Sample Rate" to every 10 minutes. Ignore any complaints about the age of the almanac.

{__} Open up the "Azimuth" window (it's under "Graphs). What causes the jumps of some satellite paths? _____

{__} Open up "Elevation". You could use this information to pick a time when a cluster of satellites was high in the sky. Change the time period by picking "Mag" (for magnify) from the "View" dropdown menu. Magnify the time scale again.

{__} Pick a point on the time scale towards the right end, noting the time you choose. Click the left mouse button. Type the character "p". Note that the time scale shifts so that the time you picked is towards the center of the time scale.

{__} Open up "SkyPlot", again under "Graphs". You see the track of each satellite in the time period shown at the bottom. If you chose a location in the U.S. notice that the north area of the plot, which is centered on the geographic location you chose, has a dearth of satellites.

{__} If your computer setup allows you to print graphics, choose "Print Graph" from the "File" menu.

{__} Under "Graphs" choose "Tile 2 Column". If you haven't closed any windows you should see the four graphs on the screen, with identical time periods.

{__} Choose "Redraw" under "View". This will restore the twenty-four hour time scale. And things will look pretty cluttered.

{__} Click the title bar in any window except the active one. It will become the active window, and will fill the screen.

{__} Under "View" pick "Close". Under "Graphs" pick "Close All".

Quick Plan can also provide tabular reports.

{__} Choose "Report Type" under "Options". You want the report that shows changes in constellations.

{__} Again under "Options", show the report. Use the "File" menu to print it. Upon examination, you can see that this would be a pretty effective planning tool, complementing the graphs. Compare this report with the one you printed from GEO-PC.

Dismiss the report by clicking once at the left of the report title bar and picking "Close". (It is easy to accidentally close Quick Plan completely here, so be careful.)

{__} Just for illustration, change the "Elevation Mask" to 40 degrees. Examine the same report and note the few times you would have enough satellites to collect data. After examining the table, choose the graph which shows the number of satellites and the resulting PDOP. It looks like those satellites between 15 and 40 degrees are vital to position finding.

{__} Make a sky plot. Note the broken red line between the two innermost circles. That is the 45 degree line. While all satellite tracks are shown, only the ones within that circle would be considered by a GPS receiver with the elevation mask set to 45 degrees.

{__} Change the elevation mask back to 15 degrees. Close all graphs and tables.

{__} You can see the power of Quick Plan. It's also fun for those who like geography. For example:

{__} Pick "Edit Point" under "Session". Pick "Cities". Find a city near you and select it. (If Quick Plan objects that it can't find the right file, exit the program completely and re-enter it. Pick "Cities" when you have a choice.)

{__} Get back to the "Edit Point" screen. Pick "World Map" to get a Mercator projection of the Earth. Move the pointer around the screen (city names will appear at the bottom; graticule coordinates show up at the top) to find a city near you. Click on it. Magnify. Again. Again.

{__} Pick "Globe". Click to free up the pointer. Find the city in which you would most like to take GPS data (or just visit). You may have to turn the globe left or right, up or down. Neat, huh?

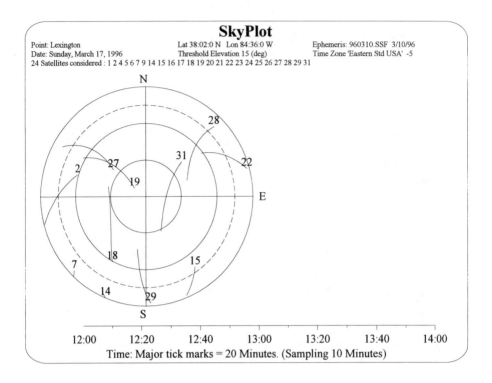

**SkyPlot**

Point: Lexington
Date: Sunday, March 17, 1996
24 Satellites considered : 1 2 4 5 6 7 9 14 15 16 17 18 19 20 21 22 23 24 25 26 27 28 29 31

Lat 38:02:0 N  Lon 84:36:0 W
Threshold Elevation 15 (deg)

Ephemeris: 960310.SSF 3/10/96
Time Zone 'Eastern Std USA' -5

Time: Major tick marks = 20 Minutes. (Sampling 10 Minutes)

Fig. 7-9 — Globe image from Quick Plan for selecting mission location

{__} "OK" the screen and leave Quick Plan. Or continue to experiment.

# Index